# AN INTRODUCTION TO HIGHWAY LAW

# An Introduction to Highway Law

**Michael Orlik,** *Solicitor*
*Partner, Needham & James, Birmingham*

**Shaw & Sons**

*Published by*
Shaw & Sons Limited
Shaway House
21 Bourne Park
Bourne Road
Crayford
Kent DA1 4BZ


© Shaw & Sons Limited 1993

First published 1993

ISBN 0 7219 1330 X



A CIP catalogue record for this book is available from the
British Library

*Printed in Great Britain by*
Bell and Bain Ltd., Glasgow

# Contents

# Foreword

The last edition of Pratt and Mackenzie's *Law of Highways*, the twenty-first, was published in 1967, the year in which I entered local government in articles to the Clerk of the West Sussex County Council. Pratt and Mackenzie contained an excellent introduction to highway law of about one hundred and sixty pages, before setting out all the statutes concerning highways, with annotations. Since that time there has been no similar work explaining the principles on which our highway law is based. In addition, the Road Traffic Acts, despite their importance in determining the way in which we use highways, have usually been covered in separate publications.

I hope, therefore, that this Introduction to Highway Law will be of interest not only to lawyers and highway engineers, but also to the organisations who look after the interests of users of the highways, such as the motoring organisations, the cyclists and the ramblers; to the police and magistracy who have to enforce the legislation, and to those in the insurance industry, who have to reach decisions on liability in the event of accidents on the highway, if litigation is to be avoided.

I have deliberately not burdened the text and footnotes with detailed citations of cases, but the reader will find a table of cases giving the relevant Law Reports at the front of the book.

I wish to record my appreciation to the County Solicitors of the County Councils of Cornwall, Cheshire, Durham, Hertfordshire, Kent and Wiltshire, who have sent me transcripts of judgments which have either been unreported, or reported only briefly in *The Times*. I wish also to thank the Municipal Mutual Insurance branch office at Bradford for letting me have a transcript of *Mills v. Barnsley Metropolitan Borough Council*.

The inspiration for this book came from my father-in-law, Roy Pitts, former Deputy County Surveyor of the Cornwall County Council. In addition, he gave me the benefit of his great practical experience as a highway engineer and encouraged and supported me throughout the writing of the book, as well as reading and commenting on the whole of the manuscript.

Mr. William Evans, an old colleague from West Sussex County Council, and now the University Secretary of the University of the West of England, and Mr. Colin Cooper, Chairman of Aspen Associates Limited, Consulting Engineers, together with his colleague, Mr. Keith Kennett, have helped me enormously by reading through the manuscript and making detailed comments and suggestions. Geoffrey Lamb, the Director of Highways and Transportation of Surrey County Council, also kindly read Chapter 10 on Traffic Regulation and made a number of helpful suggestions for improvements. I record my gratitude to all of them here. Without their aid, there would have been many more mistakes. For those that remain, and for the opinions expressed in the book, I must bear the responsibility.

Some of the cases which have determined the principles on which our highway law is based were heard in the 19th century, or even earlier. I have been fortunate in being able to use the library of the Birmingham Law Society, which has most of the old Reports, and I am grateful to the Librarians, Mr. S.P. Lahiri and Mr. J.D. Norton, who have given me much help in finding the older cases.

Mr. Crispin Williams, Managing Editor of Shaw & Sons, has been patient and encouraging at all times, and I wish to record my thanks to him also.

My partners in Needham & James have been generous in placing the resources of the firm at my disposal in preparing my manuscript. Three secretaries, Sylvia Badsey, Gail Richards and Susan Leech, have helped me with typing. I also wish to thank them, and my partners, for their constant support and interest.

Above all, it is a great pleasure to be able to record my appreciation to my wife, Susan, who has cheerfully accepted the sacrifice of a large part of our weekends for over a year to get this book written, and has also read through each chapter when it has been completed and made suggestions for improvements. If the book is intelligible as well as, I hope, accurate this is due in no small part to her clarity of thought and her insistence that the book should be understandable to a wider readership than to lawyers alone.

*Michael Orlik*

# Explanatory Note – Local Government and the Courts

Elected local bodies, raising taxation locally through rates, have been responsible for the maintenance and repair of highways for over three centuries. Unfortunately, there have been many changes in the organisation of local government over that period, and the reader may find some of the terminology and the names of the Authorities involved in the early cases puzzling.

I have described at the beginning of Chapter 3 the Authorities which currently have responsibility for the maintenance of highways. With further reorganisation of local government announced by the present Government, even that information may not be accurate in four or five years' time.

It may be helpful to explain that up to the middle of the 19th century, Parishes, Vestries and the old Charter Boroughs were responsible for the repair of highways. From the middle of the 19th century, a succession of Public Health and Highways Acts transferred responsibility for highways away from Parishes and Vestries to urban and rural District Councils. In this century, the County Councils have taken over responsibility for highways, except in the large conurbations where the London Borough Councils and the Metropolitan Borough Councils of cities like Birmingham, Leeds and Manchester have the responsibility for maintaining the highways.

So far as the Courts are concerned, some of the cases mentioned in the text have been civil actions for trespass, negligence and nuisance. Others have been criminal prosecutions for offences such as obstruction of the highway, or have related to matters where the Magistrates have been given jurisdiction, for example private street works or the stopping up of highways. The civil cases are heard in the first instance by a High Court Judge or a County Court Judge, from whom there is a right of appeal to the Court of Appeal. There is a further right of appeal from the Court of Appeal to the House of Lords.

Most of the prosecutions for offences have been dealt with by the local Magistrates, who are the Justices of the Peace. An appeal from their decisions lay to Quarter Sessions, and following the enactment of the Courts Act 1971, to the Crown Court. Alternatively, a party dissatisfied with the decision of the Magistrates on a point of law could ask them to state a case for the opinion of the Divisional Court. This Court consists of at least two, and sometimes three, High Court Judges. There is a further right of appeal from the Divisional Court to the Court of Appeal, and thence to the House of Lords.

Finally, if a Secretary of State or a Council exceeds his or its powers, or does not reach decisions in accordance with the relevant enactments, a party aggrieved by the decision can bring an action for judicial review before the High Court. These cases are reported as *R. v. ... ex parte* (at the suit of) ... . Examples are *R. v. The London Boroughs Transport Committee ex parte Freight Transport Association Limited* mentioned on page 159, and *R. v. Secretary of State for the Environment ex parte Riley*, mentioned on page 179. The applicant for judicial review asks the Court to quash the order made by the Highway Authority. The Court will not substitute its own decision. If the application is successful, and the order is quashed, the Authority making the decision will have to reconsider the position in accordance with the principles laid down by the High Court when giving its reasons for quashing the original decision.

*M.O.*

# Table of Statutes

# Table of Statutory Instruments

# Table of Cases

Chapter 1

# What is a Highway?

It is difficult to imagine a society without roads. Even in prehistoric times well-worn tracks developed across land, some of which are still shown on Ordnance Survey Maps today, such as the Icknield Way and the Ridgeway. The Romans were, of course, well known for their road building skills. Without roads it was not possible to travel between one community and another or to trade goods. The first postal services could not have been developed without roads. Once roads came into being they were also seen as an easy way of getting rid of foul and surface water and up till the last century they also served as open sewers. Today a whole network of services such as water, gas, electricity and cable television are laid under them and without roads these services could not be brought to everybody's home.

Inevitably it has become essential to develop a body of law to regulate who can use roads, for what purpose and in what manner. When the householder steps from his drive into the road he is stepping from land governed by one set of rules on to land enjoying a totally different status. Who owns the road, who has to maintain the road, who is liable for accidents caused by pot-holes in the road or trees falling across the road, can traders erect stalls in the road, is there a right to hold public demonstrations in the road, who can dig up the road and lay services in it?

It will be the object of this book to try to provide answers to these questions. The Highways Act 1980 contained 345 sections and 25 Schedules when it was passed, but even if you read through all 400 pages in Halsbury's Statutes where the Act is printed, you will not find complete answers to all these questions. There are other Acts, such as the Wildlife and Countryside Act 1981 and the New Roads and Street Works Act 1991, relating to aspects of highway law. Some matters are not covered by statute at all but have been decided

1

in the common law Courts over the centuries. Even if there is statutory provision covering a point on which a question has arisen, that statutory provision may be open to different interpretations and a body of case law may have developed on it. Sometimes the Act will say that its provisions, for example in dealing with dedication of a highway by user (section 31) or in dealing with obstructions (section 333), are not to affect any presumption or right under any rule of law which has evolved through the Courts in decided cases.

### Definition of a highway

It will be useful to start with some definitions and to concentrate first on what is a highway. The Highways Act 1980 is unhelpful and simply says that a "highway" means the whole or a part of a highway other than a ferry or waterway. This definition says nothing about the rights to use a highway, for which it is necessary to look at the decisions of the common law Courts over the last two centuries and even earlier. From those decisions it emerges that a highway is a route which all persons, rich or poor, can use to pass and repass along as often and whenever they wish without let or hindrance and without charge. At common law a highway cannot be dedicated to the use of the public subject to a right to take tolls. Mediaeval Royal Charters and modern statutes such as the Mersey Tunnel Acts, the Severn Bridge Tolls Act 1965, and the New Roads and Street Works Act 1991, have allowed tolls to be charged in a few cases. However, tolls are something of an anomaly in the concept of a highway which has for centuries been regarded as a way open to all subjects of the Crown. Since a highway is open to all, it is both confusing and wrong to refer to a *public* highway. There is no such thing as a private highway.

### Bridges

As regards bridges, section 328(2) of the Highways Act 1980 says:

> "Where a highway passes over a bridge or through a tunnel, that bridge or tunnel is to be taken for the purposes of this Act to be a part of the highway."

The definition of bridge in the next section, section 329, is as follows:

> "'Bridge' does not include a culvert but, save as aforesaid, means a bridge or viaduct which is part of a highway, and includes the abutments and any other part of a bridge but not the highway carried thereby."

The first definition makes it clear that where there is a highway at either end of a bridge or tunnel then the bridge or tunnel can be safely treated as part of the highway. In the second definition the Parliamentary draftsman has made it clear that a bridge extends beyond the highway. It includes abutments and other parts of the structure which do not form a part of the highway over which the public have a right of way.

### Cul-de-sacs or dead ends

It used to be thought that an essential characteristic of a highway was that it should begin and end in another highway. However, it has now been decided in a number of cases that dead-ends or cul-de-sacs can be highways.[1] Dedication of a road which does not lead to another highway is unlikely to be inferred merely from use by the public. However, if there is some kind of attraction at the far end, such as a beauty spot, then this may be sufficient to justify the conclusion that a highway has been created.[2]

A highway must follow a defined route on the ground. There is no right for the public at large to wander at will over open land. Public rights of access for air and exercise over certain commons granted by the Law of Property Act 1925 do not create highways. If there is no regular way and people merely go where they like, then there is no highway. However, the use of a way, such as an esplanade, for strolling up and down for pleasure, is not inconsistent with dedication as a highway providing the way is clearly defined on the ground.[3]

---

1  *Williams-Ellis v. Cobb* 1935.

2  *Roberts v. Webster* 1967.

3  *Sandgate U.D.C. v. Kent County Council* 1898.

## Meaning of 'road'

The Highways Act does not contain a definition of 'road'. Ways, roads or footpaths can be either public or private and it is meaningful to refer to either a public or a private road, path or way. The Road Traffic Act 1988 says that 'road' means "any highway *and any other road* to which the public has access, and includes bridges over which a road passes". Thus traffic offences, such as driving without due care and attention, or under the influence of alcohol, can be committed on roads which are not highways.[1] Equally, Traffic Regulation Orders can be made for roads to which the public has access even though they are not highways. Therefore, parking restrictions can be imposed and the presence of yellow lines on a road, or speed limit signs, is no proof that the road is a highway.

## Meaning of 'street'

The term 'street' is defined in the Highways Act 1980 as having the same meaning as in Part III of the New Roads and Street Works Act 1991. This provides that a 'street' means:

"... the whole or any part of any of the following, irrespective of whether it is a thoroughfare:–

(a) any highway, road, lane, footway, alley or passage,

(b) any square or court, and

(c) any land laid out as a way whether it is for the time being formed as a way or not."

This is similar to the definition of a road and again means that a street may or may not be a highway.

Part XI of the 1980 Act contains lengthy provisions for making up private streets under the Private Streets Works Code. Some streets may already be highways but are not maintainable at the public expense by the highway authority. They are often referred to as unadopted highways or unadopted streets or privately maintainable highways. It was the intention of the Parliamentary draftsman that

---

1 *Adams v. Metropolitan Police Commissioner* 1979, discussed further at end of Chapter 10.

it should be quite clear that a street, whether or not it is a highway, can still be made up at the frontager's expense if it is not publicly maintainable at the time when it is made up. Making up private streets will be discussed in Chapter 8.

## Unadopted highways

Whether a road is maintainable at the public expense or not is irrelevant to whether it is a highway or not. If the road has been dedicated as a right of way for the public, it is a highway and every subject of the Crown has the same right of passage along it, as along a highway maintained at the public expense by the Department of Transport or the local Council.

It is just as much an offence to obstruct such an unadopted highway as any other highway. Residents often think, quite erroneously, that, as they pay for the upkeep of 'their road', they can obstruct it, for example by putting down ramps to slow down traffic or by erecting locked gates: they cannot and charges of obstructing the highway or depositing materials in the highway could be brought against them.

## Classification of highways: footpaths, bridleways and carriageways

Highways can be open to all forms of traffic; for example the old cartway and the modern carriageway. However, some highways are only open to persons on foot or on horseback and are commonly called bridleways while other highways are only open to persons travelling on foot and are usually called footpaths. A highway means at least a right of way on foot. Pedestrians, horseriders and drivers of wheeled vehicles, such as carts, chariots and carriages, have all been well-known features of travel since time immemorial. The common law has, therefore, recognised three classes of highways; the road for all purposes, the bridleway for horse riders and pedestrians, and the footpath for pedestrians only.

## Special roads, motorways and trunk roads

There is an important modern exception to this ancient common law three-fold classification. A Highway Authority can be authorised by a Scheme under section 16 of the 1980 Act to construct a special road for the use of any class of traffic prescribed in the Scheme. The Secretary of State for Transport is the Highway Authority for trunk roads and it is under this power that the Secretary of State makes Schemes authorising himself to build motorways. The word 'motorway' does not appear anywhere in the 1980 Act. Section 329 of the Act provides that a special road "means a highway" but in the case of a motorway it is a highway which pedestrians, horse riders and cyclists cannot use. They are not included in the classes of traffic prescribed in the Scheme. A local Highway Authority can also be authorised to build a special road by a Scheme made by it and confirmed by the Secretary of State. The M32 connecting the M4 to the centre of Bristol is such a special road.

To find a definition of motorway it is necessary to seek out The Motorway Traffic Regulations (Statutory Instrument No. 1163 made in 1982). Regulation 3(1)(f) says that motorway means any road to which the Regulations apply by virtue of Regulation 4. That Regulation says that the Regulations apply to every special road which can only be used by traffic of classes I and II in Schedule 4 to the 1980 Act. These classes are restricted to motor vehicles. This is hardly an easy route to find the definition of motorway.

Trunk road is an expression which first appeared in the Trunk Roads Act 1936. Before that Act the local Highway Authorities were responsible for all roads. The Act empowered the Minister for Transport to designate roads of national importance as trunk roads to be vested in and maintained by him. This power is now contained in section 10 of the 1980 Act. Under section 329 of the 1980 Act, trunk road means a highway which is a trunk road by virtue of section 10 or section 19 of the 1980 Act.

It would be attractive to distinguish a trunk road from a motorway, both of which are maintained by the Secretary of State for Transport, by saying that trunk roads can be used by pedestrians, cyclists and horseriders whilst motorways cannot. This is probably one way in which many people would distinguish the two but unfortunately it is not strictly correct. Under section 19 of the 1980 Act, special roads provided by the Secretary of State in a special road scheme become trunk roads. Motorways are therefore trunk roads. It follows that a trunk road can vary between a single carriageway with many intersections and a dual carriageway with three lanes in each direction and grade separated junctions.

The important point for the purposes of this chapter is that motorways, special roads and trunk roads are all highways. It might make the position clearer if the Government in the next Highways Act actually defines motorway in the Act, distinguishes it from trunk road and stops using the phrase special road which raises more questions than it answers – for instance why is it special – presumably because it is restricted to certain classes of traffic.

### New roads – concession agreements

As mentioned earlier in this chapter, Parliament can authorise the taking of tolls on a highway. In 1991 the Government hoped to be able to arrange for new roads to be built by the private sector in consideration for the company building the road being able to levy a toll on motorists for its use. It, therefore, provided in the New Roads and Street Works Act, enacted in that year, that the Highway Authority could enter into a Concession Agreement with any person for the construction of a new road. In return for undertaking obligations with respect to the design, construction, maintenance and improvement of the road, the person or company entering into the Agreement, whom the Act calls 'the Concessionaire' would be able to enjoy the right to charge tolls for the use of the road. Any road built in this way will be a special road within the meaning of the Highways Act 1980. The Concessionaire can charge what tolls it likes, except in the case of a crossing of navigable waters more than one hundred metres wide, such as a river estuary, when the Secretary of State will specify a maximum rate of toll.

A special road in the 1980 Act means a highway and under section 36 of the Highways Act 1980, a special road is maintainable at the public expense. Subject only to the payment of the appropriate toll, any member of the public has a right of passage along the road built by the Concessionaire exactly as on any other highway. Under the Concession Agreement, the Highway Authority, be it the Secretary of State for Transport or a local Highway Authority, can pass some or all of its maintenance responsibilities to the Concessionaire. The general law regarding highways, as set out in the rest of this book, therefore applies to these new roads as much as to roads built by the Highway Authorities operating within the public sector. The powers and duties of the Highway Authority, described later in this book, will be applicable but the Agreement with the Concessionaire will set out how these are to be divided between the Authority and the Concessionaire. The Authority cannot delegate powers to make Compulsory Purchase Orders to acquire land, nor powers to make Traffic Regulation Orders (see Chapter 10) other than temporary orders not lasting more than a few days. As these new roads will be special roads, the scheme establishing them will make provision for prescribed classes of traffic to use them and pedestrians and cyclists are unlikely to be allowed on them. It is possible that some of these roads will be created as lorry-only roads.

So far, no new roads have been built under the 1991 Act.

### Private ways

The Highways Act includes in its interpretation section meanings of the terms footpath, bridleway and carriageway and describes these terms respectively as rights of way for the public on foot, on horseback and with vehicles. Whilst this is helpful when considering the provisions in that statute, outside the Act these terms can still be correctly used in everyday language to describe private paths and drives though, at least so far as bridleways are concerned, most people nowadays will understand this term to mean a right of way on horseback for the public at large and not a private path.

### Footpaths and footways: pavements

The 1980 Act defines a footpath as a highway over which the public have a right of way on foot only, not being a footway. It defines a footway as meaning "a way comprised in a highway, which also comprises a carriageway, being a way over which the public have a right of way on foot only". In towns this means a path beside a road known to everyone as a pavement. Pavement however is not defined in the statutes and would not be the correct term for a surfaced path running beside a road in the countryside.

## Cycles

The bicycle was not invented until a century ago when the common law on the classification of highways was well developed. Cyclists have only, therefore, been able to use the highways dedicated for wheeled traffic until quite recently when section 30 of the Countryside Act 1968 enabled cyclists to ride on bridleways. A 'cycle track' is a recent statutory term now defined in the Highways Act 1980 as a way constituting or comprised in a highway over which the public have a right of way on pedal cycles and may have a right of way on foot.[1]

Section 72 of the Highway Act 1835 is still in force and it is an offence under that section for any person to "ride upon any footpath" set apart for the use of pedestrians by the side of a road. Such a footpath, as explained above, is now defined as a footway. Although the bicycle was invented after 1835, cyclists can be prosecuted under this section for riding their cycles on footways. The section only applies to paths at the side of a road. Cyclists cannot be prosecuted for riding on footpaths, whether surfaced or unsurfaced, away from roads unless a local byelaw or traffic regulation order (see Chapter 10) has been made forbidding cycling on the path in question.

## Byways

The Wildlife and Countryside Act 1981 has introduced two new classifications. It provides that a byway open to all traffic means a highway over which the public have a right of way for vehicular and all other kinds of traffic but which is used by the public mainly for the purpose for which footpaths and bridleways are so used. In plain English, it would be simpler, if less accurate, to say that a byway is a bridleway along which motorists and motor-cyclists can also pass without committing an offence. Under the National Parks and Access to the Countryside Act 1949, these ways were defined as "roads used as public paths" and they have often colloquially been referred to as "green lanes".

---

1  See also Cycle Tracks Act 1984.

## Public paths

The 1981 Act also says that a 'public path' means a highway being either a footpath or a bridleway. This is a convenient definition for the purposes of the provisions in that Act relating to Definitive Maps of rights of way, but it should not be assumed that 'public path' where the phrase appears in other Acts or deeds and documents necessarily has this meaning. The Act is comparatively recent and much of the law relating to highways still depends on decisions of the old common law Courts.

## Walkways

In recent years, in order to cope with large, covered shopping complexes, a new hybrid creature has been introduced called a walkway, the creation of which is now governed by section 35 of the 1980 Act. The owner of a building can enter into an Agreement with the Highway Authority or with the District Council to provide ways over, through or under parts of the building and "for the dedication by that person of those ways as footpaths subject to such limitations and conditions, if any, affecting the public right of way thereover as may be specified in the Agreement". A footpath created in pursuance of such an Agreement is referred to as a "walkway". Whilst the Act uses the word "dedication", the Agreement may provide for the termination of the right of the public to use such a walkway and will usually provide that the maintenance of the walkway is to be carried out by the owner of the building, rather than by the Highway Authority. The attraction to the owners or managers of shopping centres in entering into these Agreements is that the local authority can then make by-laws regulating the conduct of persons using the walkways and the times at which the walkways may be closed to the public so that the ways can be closed and locked at night-time. The existence of the by-laws also makes it easier for the police to work in the shopping centres and to deal with disorderly behaviour.

A walkway certainly does not conform to the common law criteria for a highway since it can be closed at night or taken out of public use permanently under the terms of the agreement entered into with the owners. It is an open question therefore whether a walkway can properly be described as a highway. An indication of the difficulty in categorising a walkway is the reference in section 115A(5) of the 1980 Act in dealing with Local Act walkways to a "highway only for certain purposes". There is also power under sub-section (11) of section 35 to make regulations preventing any enactment relating to highways from applying to walkways, or for adapting any such enactment.

The present Regulations were made in 1973.[1] Schedule 1 of the Regulations lists a large number of sections of the 1980 Act which do not apply to walkways; Schedule 2 lists sections in the 1980 Act which are not to be affected by the provisions of the walkway agreement with the building owner; Schedule 3 lists a number of sections which allow the Highway Authority to carry out works in any highway and provides that those powers in the case of a walkway are not to be exercised without the consent of the building owner. Accordingly, anyone wanting to know whether a particular section of the 1980 Act does, or does not, apply to a walkway must look at the Schedules to the 1973 Regulations. So far as members of the public are concerned, as a general rule, the provisions in the 1980 Act prohibiting interference with highways discussed in Chapter 5 apply to walkways so that, for example, a person could be prosecuted for obstructing a walkway in the same way as any other highway.

## Summary of definitions

It may be helpful (but inevitably not completely accurate) to summarise below the definitions discussed in this chapter:

*Highway:*
A right of way for the public on foot. It may also be a right of way for persons on horseback or driving vehicles.

*Public Path:*
A highway which is a right of way on foot and on horseback.

---

1  S.I. 1973 No. 686.

*Byway Open to All Traffic:*
A highway over which the public have a right of way for
all purposes but which is used mainly as a footpath and
bridleway.

*Bridge:*
Part of the highway, if there is a highway at either end.

*Road:* A way for vehicles.

*Street:* A way for vehicles.

*Bridleway:* A way for riders on horseback.

*Footpath:* A way for pedestrians.

*The four terms immediately above describe ways by
reference to the sort of traffic which can physically use
them. These ways may be highways over which the
public have a right of way, or they can be private.*

*Cycle Track:*
A highway over which the public have a right of way on
pedal cycles with or without a right of way on foot.

*Walkway:*
A way dedicated as a right of way for the use of the public
on foot, subject to limitations, e.g. closure at night time.

Although the first three terms in this list describe rights of way for
the public, they are not necessarily maintainable at the public
expense, either by the Department of Transport or by a local
Highway Authority. Liability for maintenance of highways will be
dealt with in Chapter 3. On the other hand, if a road or footpath is
maintained at public expense, that is strong evidence that the way
in question has been dedicated as a highway to the public and
accepted by the Highway Authority. Dedication, acceptance and the
creation of new highways will be addressed in the next chapter.

# Chapter 2

# Creation of Highways

## Dedication by owner

Highways come into existence through dedication of a right of way to the public by a landowner and acceptance by the public of that dedication. Dedication by the owner or by a previous owner is usually a matter of inference. Although there are thousands of roads and paths, there have been relatively few explicit acts of dedication, or, if there were, records have not survived. There is no standard form of dedication which owners are required to follow today if they wish to dedicate a way as a highway. A short statement signed by an owner dedicating a highway over a piece of land shown on an attached plan is sufficient. Use of the way by the public is sufficient evidence of acceptance.

Whilst use by the public is usually conclusive evidence of acceptance of a way as a highway, use by the public is not necessarily evidence of an intention on the part of the landowner to dedicate the way as a highway. In the absence of a written form of dedication, the question of dedication is one of fact to be determined on all the available evidence, of which public use is only one part. It is often a difficult, and sometimes hotly disputed, issue.

Most roads and paths have been used for so long that it is often not possible to find evidence of when the way was first regarded as a highway, who was the landowner at that time and what his intentions may have been with regard to dedication. Consequently the history and nature of the use of the way are always important. The use must be *as of right* and *without interruption* in order to prove an intention to dedicate.

## Use as of right

Use by permission of the landowner is not use as of right. Use of a private way by servants or tradespeople or the postman to reach the owner's house does not indicate an intention to dedicate the way to the public at large. Similarly, use of a forecourt of a shop or public house adjoining the highway is usually restricted to customers and by permission of the owner of the premises. It has been held in decided cases that such use makes it more difficult to prove dedication because it is not easy for the owner to distinguish those people whom he has permitted to be on the way, from members of the general public trespassing, whom he, or his agents, would otherwise accost and turn back.[1] Permission by the owner limited to the inhabitants of the parish to use the way, or permission for anyone to reach a church does not amount to an intention to dedicate a right of way to the public at large.

The reputation of the way locally can be important and evidence from local people as to whether they thought they were using the way with the permission of the landowner or whether they had always regarded the way as a public right of way is admissible. Old public documents such as Inclosure Awards and tithe maps are also admissible. Inclosure Awards will, for instance, quite often refer specifically to a "private carriage road". Again, this is not necessarily conclusive because the owner may, at some time after the Award was made, have dedicated a right of way on foot or with vehicles to the public over the carriage road. The location of the way can also be significant. If it is obviously a useful direct route between two existing highways which have been there for a long time then that is suggestive of the connecting route also being a highway. It is often helpful to look at the route on the ground.

The route must be capable of being used openly and without having to circumvent obstacles. Reaching the route via a hole in the hedge or by climbing over a locked gate, or by squeezing round the pillars of a gate is evidence that the way is not being used as of right. Unlocked gates are capable of interpretation either way. They are not necessarily evidence of no intention to dedicate, because they may have been erected by the owner to prevent cattle straying, not

---

1  *R. v. Bradfield Inhabitants* 1874; *Holloway v. Egham U.D.C.* 1908.

to keep the public out. On the other hand they may have been erected to deny access and may rebut any intention to dedicate.

## Use without interruption

As regards interruption, it will be sufficient for an owner to close the way for one day a year to negate any intention to dedicate. Alternatively, any action by the owner bringing into question the existence of a right of way is powerful evidence of a refusal to dedicate. An owner cannot police his land 24 hours a day and the burden of having a public right of way across his land is a heavy one. The Courts have, therefore, tended to place much more weight on evidence showing that on a few occasions the owner has specifically challenged the right of way than on evidence showing that on many other occasions the public have been able to use the right of way without interference or challenge.[1]

## Evidence of maintenance of the way

Before 1st August 1835, dedication and acceptance had important financial consequences because the inhabitants of the parish became liable to maintain the highway immediately it had been accepted by public use. Since that date highways do not become publicly maintainable simply by dedication and acceptance. Thus it is not possible for, say, a farmer to throw open to the public a road leading through his farm in the hope that the Highway Authority will then have to bear the expense of maintaining the road so that the milk lorry can use it to reach the dairy. The dedication by the farmer and use by the public will make the road into a highway but not a publicly maintainable one.[2]

The absence of any public expenditure on repair or maintenance is not, therefore, good evidence that the way is a private one. This is because since 1836 there can be dedication and acceptance of a public right of passage without the highway becoming maintainable at public expense. On the other hand, the fact that a road has been

---

1 *Chinnock v. Hartley Wintney R.D.C.* 1899.

2 See the next chapter for an explanation of what is a maintainable highway .

repaired or maintained at the public expense is strong evidence that it is a highway.

## The presumption of dedication and the twenty year rule

Until the passing of the Rights of Way Act in 1932, now incorporated in section 31 of the Highways Act 1980, the persons arguing for the existence of a right of way, as well as having to prove use as of right and without interruption, also had to demonstrate that there was a landowner with the power to dedicate a right of way at the time they claimed the path came into existence. As one Judge put it in describing the state of the law before 1932:

> "Above all the other difficulties, the Tribunal had solemnly to infer as an actual fact that somebody or other had in fact dedicated. It was often a pure legal fiction, and yet put on the affirmant of the public right an artificial onus which was often fatal to his success."[1]

Before 1932, a tenant, even a tenant for life, had no power to dedicate a right of way. If an owner disputing the existence of a right of way could show that at the time when it was alleged to have been dedicated some time previously, the owner was a tenant for life or a child or somebody living abroad, he would often be successful in persuading the Court that no public right of way could have been dedicated.

Section 31(1) of the Highways Act 1980, following the 1932 Act, now provides:

> "Where a way over any land, other than a way of such a character that use of it by the public could not give rise at common law to any presumption of dedication, has been actually enjoyed by the public as of right and without interruption for a full period of 20 years, the way is to be deemed to have been dedicated as a highway unless there is sufficient evidence that there was no intention during that period to dedicate it."

---

1 *Jones v. Bates* 1938.

16

This period of twenty years' use has become familiar to all those concerned with mapping rights of way and arguing for or against their existence. It should be noted, however, that section 31(9) says:

"Nothing in this section operates to prevent the dedication of a way as a highway being presumed on proof of use for any less period than 20 years, or being presumed or proved immediately before the commencement of this Act."

At common law there was and is no fixed minimum period of use required to establish dedication. If a clear intention to dedicate can be shown, quite short periods of use of only a few years have been held to be sufficient.[1]

The effect of the legislation has therefore been to remove the difficulties about proving that there was a landowner capable of dedicating the way as a right of way some time in the past but the statutory provisions have not changed the other principles laid down by the common law Courts and discussed above. It is still necessary to show use as of right and without lawful interruption.

A landowner may still be able to show that a right of way could not have been dedicated because the land was held by a tenant for a term of years[2] at the time of the claimed dedication or was held for charitable or statutory purposes inconsistent with the dedication of a right of way.[3] However the onus is now on him, once the twenty years' use as of right and without interruption has been proved, to show that the right of way could not have been dedicated. The claimant no longer has to show that it could have been dedicated.

Although section 31(11) lays down that land includes land covered by water, the House of Lords held in 1991 that a river does not become a highway after twenty years' uninterrupted navigation by

---

1 *North London Railway Co. v. St. Mary Islington Vestry* 1872.

2 Under section 56 of the Settled Land Act 1925 tenants for life now have power to dedicate land as a highway.

3 *R. v. Leake Inhabitants* 1833; *Paterson v. St. Andrew's Magistrates* 1881.

the public. Parliament had intended the Act to apply to a way over water, such as a ford, but not to a way along the length of a river.[1]

### Evidence that no dedication intended

It is still possible for a landowner to rebut the presumption raised by twenty years' use by proving that there was no intention during that period to dedicate a public right of way. Section 31 provides an owner with means of establishing that he has no intention of making such a dedication. He can either erect and maintain a notice saying that there is no right of way or, if the notice is torn down or defaced, he can deposit with the Highway Authority a map and statement showing which ways over his land he admits to have been dedicated as highways. If he lodges Statutory Declarations every six years that no additional ways, other than those shown on the map or mentioned in the Statutory Declarations, have been dedicated, then, in the absence of proof of contrary intention, this would be sufficient evidence that the owner and his successors in title have had no intention of dedicating any additional ways.

### Limited dedications

Dedication of a highway must be for all time and cannot be for a limited period. Similarly, it cannot be for a limited class of persons or only a section of the public. It is, however, possible to dedicate land as a highway subject to other limitations. A highway can be dedicated for use on foot, on foot or horseback, or for vehicular traffic. An owner in dedicating his land can stipulate the classes of traffic to which he intends the way to be open or can limit dedication. Towpaths have often been made the subject of such dedications.[2] In *Grand Junction Canal Co. v. Petty* (1888), the Master of the Rolls said:

> "That such a dedication is to be taken to be an absolute dedication I cannot suppose. It must be, I think, a dedication to the public of the towing path, for the

---

1  *Attorney General v. Brotherton* 1991.

2  As to cases on two different lengths of towpath on the River Thames, see *Winch v. Thames Conservators* 1874 and *Thames Conservators v. Kent* 1918.

purpose of such use as a footpath as will not interfere with its ordinary use as a towing-path by the company. I do not think that the public walking on the towing-path are entitled to say that the towing must be regulated with reference to their convenience. The public in accepting the dedication must be taken to accept it as a limited dedication, and cannot set up a right to prevent or limit the use of the towing-path by the company. If the horse or the tow rope and the foot passenger are in one another's way, the foot passenger must look out for himself and get out of the way."

The Wildlife and Countryside Act 1981 recognises that conditional dedications are still possible by requiring the Definitive Statement of rights of way to include particulars of limitations or conditions affecting a particular right of way (see Chapter 11).

A road can be dedicated as a highway even though it is impassable in winter. A way can also be dedicated subject to permanent obstructions such as trees, gates, stiles or projecting doorsteps. It is quite common for a highway to have been dedicated subject to existing market rights so that market stalls or fair booths can be erected for temporary periods to the obstruction of the right of way. A swingbridge can be dedicated as a highway subject to a right to interrupt traffic to allow vessels to pass. As owners are under no obligation to dedicate land for a highway, and as members of the public are under no obligation to accept a dedication, they cannot complain that the owner has dedicated an unsatisfactory or dangerous highway and they must take the way as they find it.

What an owner cannot do is to impose restrictions after dedication has taken place. Moreover, if for some time the owner has not insisted upon restrictions originally imposed, or has not exercised a right originally reserved of obstructing the highway, it is a question of fact whether or not he has abandoned the rights originally retained by him. These principles were considered in *Gloucestershire County Council v. Farrow* (1983).

In this case, the Lord of the Manor had been granted a Royal Charter in the twelfth century to hold a weekly market in a square in the middle of Stow-on-the-Wold in the Cotswolds. By the nineteenth century the square had been dedicated as a highway and in consequence there was a right of way over the square subject to the right to use the square for weekly markets. Around 1900 the market ceased to be held. In 1979 the Lord of the Manor leased his market rights to a company, of which Mr. Farrow was a Director. This company wished to revive the weekly market.

The County Council, as the Highway Authority, brought an action in the High Court seeking a declaration that, because the square was a highway, the obstruction of the square by market operations would be an unlawful obstruction of the highway. The Judge accepted the argument of the Highway Authority that, as the square had been continuously enjoyed as a highway by the public as of right for a period of over twenty years, the public had now acquired an unqualified right of way over the square by reason of section 31 of the Highways Act 1980, discussed earlier in this chapter. In the absence of sufficient evidence that there was an intention during that period of twenty years to maintain the market, the square was deemed to have been re-dedicated as a highway free from the market rights. The County Council was granted an injunction forbidding the company from going ahead with its plans to resume the holding of weekly markets in the square.

The great majority of highways are not dedicated subject to any limitations. It is, however, important to keep in mind the possibility that there has been a limited dedication in considering whether offences of interfering with a highway have occurred. There are a great many sections in the Highways Act prohibiting interference of different kinds with a highway but several of them, such as prohibition of obstructing a highway (section 137) and prohibition on depositing objects in the highway (section 148) begin:

"If, without lawful authority or excuse ... "

If, for instance, a farmer had dedicated a footpath subject to the existence of a stile, he cannot be guilty of obstructing the highway. "Without lawful authority" usually means the authority of the owner or the Highway Authority and, therefore, does not authorise other persons to create obstructions, even if there has been a limited dedication of the way.

## Creation of highway by statute

In the present century Parliament has, in addition, laid down a number of ways in which highways can be created without dedication or use. The most obvious is the power of the Highway Authority, be it the Secretary of State for Transport or the local council, to build a new road under section 24 of the 1980 Act. Any road built by a Highway Authority becomes a highway maintainable at the public expense but an understanding of some of the principles set out above about dedication is still helpful. There is no power to make a landowner dedicate a right of way across his land. In order to build a new road, the Highway Authority will have to purchase, by agreement or compulsorily, the freehold ownership of the land so that, as owners, they can build the road across the land. Moreover, after they have acquired the land, but before they have built the road and opened it to the public, there are no rights of passage over it for the general public. Highway authorities often acquire ownership of pieces of land gradually but until they have been able to build a continuous length of road between two existing highways and until they have been able to make the surface of the road or path passable for the public, the necessary dedication and acceptance of the land as a highway has not taken place.

In a case where the local County or District Council thinks that there is a need for a new public footpath or public bridleway, they do not have to acquire ownership of the land over which the path is to run. They can, under section 26 of the 1980 Act, make a Public Path Creation Order and submit it to the Secretary of State for confirmation. Although in such a case they do not have to buy the freehold of the land over which the path is to run, they will have to pay compensation to the landowner for any depreciation in the value of his land through the creation of the right of way over it.

A Parish Council may enter into a voluntary agreement under section 30 of the 1980 Act with a landowner for dedication of a highway over land in the parish. Such an agreement may contain provisions as to how the owner and the Parish Council are to deal with construction of the highway and its future maintenance because a highway created in this way will not be the responsibility of the Highway Authority to keep in repair. Section 30 is not, therefore, a widely used power.

### Agreements with builders for laying out new estates – Highways Act 1980 section 38

Perhaps the most common way of creating a new highway nowadays is the agreement made under section 38 of the 1980 Act. Where a developer is laying out a new housing estate, or a new industrial estate, or indeed any other form of development and is providing roads as part of that development, he will usually want the Highway Authority to take over the maintenance of the roads once they have been built and the development has been completed. There are a number of reasons for this. The developer or builder will not wish to continue to maintain the roads once he has constructed the buildings and sold them. The purchasers of the houses, or other buildings, will also not usually wish to be responsible for the maintenance of the new roads but will wish to be assured that the roads will be maintained in a good state in perpetuity so that they can use them comfortably. They will also wish to avoid possible future liability for street works charges if the Highway Authority at some later date decides that the roads should be made up. On new housing estates most purchasers' solicitors and most building societies will, as a matter of course, want to be satisfied that a Section 38 Agreement has been entered into by the developer and that the roads will be adopted by the local Highway Authority. It is becoming more common in the case of some commercial and industrial estates for the occupiers of the buildings not to be so concerned with adoption of the roads as highways because, for reasons of security, they may want to put gates across the roads and keep out traffic at night time. The developer and his purchasers may decide in such a case that it would be preferable to retain the estate and the roads as private and not to make any dedication of public rights of way.

An Agreement made under Section 38(6):

> "may contain such provisions as to the dedication as a

highway of any road or way to which the Agreement relates, the bearing of the expenses of the construction, maintenance or improvement of any highway, road, bridge or viaduct to which the Agreement relates and other relevant matters as the Authority making the Agreement think fit."

The Highway Authority is, therefore, in a strong position in insisting on dedication as a highway and on a high standard of construction before it agrees that the road shall be publicly maintainable. The Authority will normally insist that the agreement be supported by a bond entered into by a bank or insurance company so that if the builder fails to complete the new road, the Authority can do the work itself and recover the cost from the bondsman. The solicitor for the purchaser of a new house on a new estate will be able to ask the Authority whether the Section 38 Agreement is supported by a bond.

It is sensible for the developer to agree with the council the form of a Section 38 Agreement before he starts construction work. If, however, he does not do so, or considers that the council are insisting on unnecessarily high standards of construction, but still wishes the road to become a highway maintainable at the public expense, he can take a risk and build the road. He can then give notice that he intends to dedicate the road as a highway. If the council wish to refuse to adopt the road as a highway they will have to apply to a Magistrates' Court for an order that the proposed highway "will not be of sufficient utility to the public to justify its being maintained at the public expense" (section 37). Alternatively, the council can accept the dedication but can refuse to certify that the road has been made up in a satisfactory manner or that it has been kept in repair for twelve months after dedication or that it has been used as a highway during that period. In that event the developer can appeal to the local Magistrates, and if he can satisfy them that there is no validity in these objections and that a certificate ought to be issued, then the Magistrates may make an order that the road is to become a highway maintainable at the public expense.

### Declaration of a highway after street works

Finally, where the Highway Authority have executed street works in a private street (discussed later in Chapter 8) they may, under section 228 of the 1980 Act, give notice that the street is to become a highway. The owners of a street, which usually means the frontagers, can object to such a notice and, in that event, the Highway Authority can apply to the local Magistrates for an order overruling the objection and declaring the street to be a highway. The Magistrates have a wide discretion in such a case and again are likely to take into account how far the street will be of use to the general public.

### The width of highways

Until quite recently no Act of Parliament has laid down measurements for the width of highways. Some eighteenth and nineteenth century Inclosure Awards did specify the widths of highways to be laid out. Sometimes extremely generous widths of twenty or thirty feet were stipulated and it is unlikely that the highways were made up to those widths or became repairable by the inhabitants at large over the whole width specified. During this century, some of the statements accompanying the maps of rights of way prepared under the National Parks and Access to the Countryside Act 1949 do give in their descriptions of the paths shown on the map the width existing on the ground. In general there will not be documentary evidence of the width of a highway and, if there is, it may well not be conclusive. What matters more is what exists on the ground. That is to say, the extent of the land which appears to have been dedicated for the public to pass over and whether, in fact, the public have demonstrated acceptance of the dedication by walking or driving over the land. If the Highway Authority has maintained land at the side of the road, as well as the metalled road itself, that is strong evidence that the land is part of the highway and is maintainable at the expense of the public.

The rights of public passage and the consequential restrictions on the powers of owners to deal with their land as they see fit have meant that there have been plenty of disputes as to the width of particular highways and there is a good deal of case law. However, decided cases may not be particularly helpful in this context because the extent of the land subject to the public right of passage is usually a question of fact. It is often helpful to look at a site before trying to reach a conclusion. As well as seeing whether there have been acts of maintenance by Highway Authorities, the existence of statutory undertakers' apparatus such as telephone cables, electric cables and gas mains, can be a telltale sign. If the undertakers have obtained wayleave consents from adjoining owners to place the apparatus in, say, a verge at the side of the road, that suggests that the verge is not part of the highway. If, on the other hand, they have not obtained any wayleaves, then this suggests that they are using their statutory powers and the Public Utilities Streetworks Code to lay services in the highway without the need to obtain consents of any private owner. (This Code is discussed later in Chapter 9.)

**Fences and verges**

The existence of a metalled road may be a good indication of the extent of the highway when such a road crosses unenclosed land such as a heath or common. It is no indication of the extent of the highway in other cases. Where there are fences or ditches on both sides of the highway the public right of passage will be taken to be the extent of the whole space between the fences or ditches even though the width of the highway may be varying and unequal and even though there may be a substantial amount of land lying between the metalled road and the fence. The presumption that the fences mark the highway boundary can be rebutted by evidence. The existence of the fence must be referable to the highway. If the fence existed before the highway, or if the distance between the carriageway and fences is so great that no connection is apparent, then the intervening land will not necessarily be regarded as part of the highway. If it is found that the fences are referable to the highway there is a further presumption that all the land between the fence has been dedicated even though there is no evidence of dedication or public use but this presumption can be rebutted by acts of ownership inconsistent with any intention to dedicate the land outside the made up carriageway.

These principles were applied by the High Court in *Attorney*

*General v. Beynon* (1969). Mr. Beynon lived in a property fronting a metalled road twenty feet wide. The road also had verges on either side and on the opposite side of the road to Mr. Beynon's property was a wide verge. It was about forty feet in width between the metalled road and an ancient hedge and ditch. From around 1953 Mr. Beynon had kept a number of cars and farm vehicles on this verge, on the opposite side from his house, in connection with a business he was operating of buying and selling vehicles.

The Attorney General applied to the High Court on behalf of the Leicestershire County Council, as the Highway Authority, for a declaration that the whole of the verge between the edge of the metalled road and the hedge was part of the highway. The Judge summarised the law by saying:

> "It is clear that the mere fact that a road runs between fences, which of course includes hedges, does not *per se* give rise to any presumption. It is necessary to decide the preliminary question whether those fences were put up by reference to the highway that is to separate the adjoining closes from the highway or for some other reason. When that has been decided, then a rebuttable presumption of law arises, supplying any lack of evidence of dedication in fact, or inferred from use that the public right of passage, and therefore, the highway, extends to the whole space between the fences and is not confined to such part as may have been made up."

Applying these principles to the facts of the case, the Judge found that no evidence had been produced by Mr. Beynon, as the defendant, to rebut the presumption that the hedges had been grown with reference to the highway. As regards the second presumption as to whether it had been intended to dedicate the land between the hedge and the metalled road, Mr. Beynon was very much at a disadvantage in that he could not show ownership of any land on that side of the road. The Judge found that his actions in parking vehicles on the verge were not the acts of an owner but the conduct of a stranger taking advantage of a convenient strip of land. Thus, although the council had failed to prove any acts of public right over the verge, except close to the metalled carriageway, the actions of

Mr. Beynon were insufficient to displace the presumption that the public right of passage extended to the whole space between the hedge and the metalled road.

Where a fence bordering a highway has on the roadway side of it a ditch, there is a presumption that the ditch does not form part of the highway. Following a case in 1938[1] when an adjoining owner successfully established ownership of a ditch in these circumstances against Bedfordshire County Council, statutory power, now contained in section 101 of the 1980 Act, has been given to the Highway Authority to fill a ditch adjoining or lying near to the highway with the consent of the adjoining occupier or to pipe it without his consent, and then fill it in, subject to the payment of compensation for any damage suffered by the occupier. This avoids arguments about whether the ditch is, or is not, part of the highway.

## Gates

As explained earlier in this chapter, ways can be dedicated as highways subject to the existence of gates across them. Section 145 of the 1980 Act provides that where there is a gate of less than ten feet width across a highway consisting of a carriageway, the Highway Authority may require the owner of the gate to enlarge the gate to that width or remove it. Similarly, where there is a gate of less than five feet across a bridleway, the Authority can require the owner to enlarge the gate to that width or remove it.

## Width of rights of way – statutory measurements

In cases when the width of the highway dedicated cannot be proved, the Rights of Way Act 1990, for the first time, contains some rather more detailed provisions about highway widths. These provisions do not relate to made up carriageways; they only apply to byways and public paths. The Act provides a minimum width of one metre for a footpath crossing a field and two metres for a bridleway. As regards a footpath going around the edge of a field, the Act specifies one-and-a-half metres. If a farmer ploughs out his field, he must reinstate a footpath or bridleway to these minimum widths and must

---

1  *Hanscombe v. Bedfordshire County Council* 1938.

ensure that paths of those minimum widths are left through growing crops. There appears to be no minimum width specified for a bridleway going round the edge of a field. There is a minimum width for any other highway, i.e. byway open to all traffic, of three metres.

In the event of the farmer failing to reinstate the paths as required, the Highway Authority can carry out reinstatement work to make good the surface of the paths. The Act also specifies, for the first time, maximum widths for such paths, so that in entering on the land to carry out the work, a width no greater than 1.8 metres can be cleared for a footpath and no greater than three metres for a bridleway. The Authority can clear a width of up to five metres for any other highway which is not a made up carriageway.

### Roadside waste

In the Middle Ages when the manor was an important unit, both of land ownership and of administration, land which was not cultivated was either common land or waste of the manor. The Commons Registration Act 1965 requires both 'common land' and 'waste land of a Manor' to be registered, except where the land is part of the highway. Waste land abutting the highway is often referred to as roadside waste and section 130 of the 1980 Act requires the Highway Authority to protect the rights of the public to enjoyment of any highway and goes on to say that the Authority shall also protect 'any roadside waste' which forms part of the highway. Whether or not the waste does form part of the highway is a difficult issue. The fact that the public cannot easily walk over it does not preclude it being part of the highway, especially if the land on either side is fenced. As between the Lord of the Manor and the owner of the adjoining land there was a presumption that the waste land belonged to the adjoining owner and that he, not the Lord of the Manor, could exercise certain acts of ownership such as taking the grass. However, if the roadside waste adjoined or lay close to much larger portions of common or waste land, the Lord of the Manor might be able to establish that ownership of the waste still lay with him.

Registration of lands as common or waste lands under the 1965 Act is definitive. If the Highway Authority did not object to registration in due time and the land is shown on the register kept by the Local

Authority it cannot now be treated as part of the highway. Under section 45 of the 1980 Act, the Highway Authority may search for and dig gravel, sand, stone and other materials from any waste or common land, whether or not the land adjoins the highway. They can use these materials in repairing and maintaining highways without paying compensation unless they are excavated from inclosed land.

**Village greens**

The 1965 Act also required the registration of town and village greens. The definition of town or village green did not exclude highway land and it is therefore possible for land to be registered as a green which may also be a highway – for example those parts of a green nearest the metalled roads surrounding the green. The point can be important if the Highway Authority ever wish to widen the carriageway of the road or to surface a footway by taking land from the green. There is strong statutory protection for village greens, making it extremely difficult to carry out any kind of work on them but if the land closest to the road can be shown to be part of the highway, the Highway Authority has ample powers to widen the road.

Chapter 3

# Responsibility for Maintenance of Highways

## The creation of Highway Authorities

Up to 1835 all highways were maintainable by the inhabitants at large unless it could be shown that responsibility had attached to an individual or a corporate body by reason of tenure, enclosure or prescription. Highways Acts in the 20th century no longer use the phrase 'inhabitants at large' but refer to highways 'maintainable at the public expense'.

The duties of the inhabitants at large to maintain highways fell upon the inhabitants of each parish who were bound to repair all the highways within their area, except those which were privately maintainable. After the Highways Act 1862, parishes could combine to form a Highways Board to carry out their highway duties and since then a variety of local authorities have had responsibilities for the maintenance of highways. At the present day the situation has been much simplified. The Secretary of State for Transport is the Highway Authority for motorways and trunk roads.[1] The County Councils are the Highway Authorities for all highways in their counties except motorways and trunk roads. Outside the areas of the County Authorities, the London Borough Councils and the Metropolitan District Councils, for example Birmingham and Sheffield, are the Highway Authorities for all the roads in their districts, except again motorways and trunk roads.

---

1 As explained in Chapter 1, motorways, except in the rare cases where they are provided by a local Highway Authority, are trunk roads. However, since the non-specialist probably regards a motorway as being a rather different road from a trunk road, the phrase "motorways and trunk roads" will be used throughout this book.

## Agent Highway Authorities

Although the County Councils are the Highway Authorities for the highways in their counties, the District Councils can, if they wish, under section 42 of the 1980 Act undertake the maintenance of roads in their built-up areas, roads subject to a speed limit of thirty or forty miles per hour and all the footpaths and bridleways in their district. The County Council must reimburse to the District Council any expenses incurred by them in carrying out maintenance work. Roads which are trunk roads or classified roads are excluded from section 42 even though they may, where they pass through built up areas, be subject to speed limits.

It is simpler for a District Council which includes a substantial town, such as Leicester or Woking, to maintain all the highways in their area. In practice, therefore, section 42 powers are rarely claimed by District Councils. Instead, the County Councils voluntarily enter into Agency Agreements with District Councils under section 101 of the Local Government Act 1972 for the District Councils to maintain all the roads within their district. The County Councils will usually reserve some functions to themselves, such as the construction of major new roads or the maintenance of bridges. The powers reserved and powers delegated depend upon the terms of the individual Agency Agreements.

These comments do not apply to motorways and trunk roads. Such roads are the responsibility of the Department of Transport. They will usually employ the County Councils to act as their agents.

## Which highways are maintainable at public expense?

The Highways Act 1980, section 36, sets out which highways are to be maintainable at public expense. As regards existing highways, all highways in existence before 31st August 1835, and any highways coming into existence after that date which can be shown to have been maintained at the public expense 'at some time' are now maintainable at the public expense. Any other highway is not maintainable at the public expense unless the person dedicating it as such is able to persuade the local Magistrates that it is of sufficient utility to justify it being maintained at the public expense (section

37) or unless the Highway Authority have agreed under section 38 that the highway should be so maintainable.

Any new road, constructed by the Department of Transport or a local Highway Authority, is maintainable at the public expense.

The Highway Authority must keep a list of all the highways within their area which are maintainable at the public expense and must make that list available for public inspection. Information is also supplied to any purchaser of property, who uses the standard form of enquiries of the Local Authority and pays the prescribed search fee, as to whether the road abutting the property he is purchasing is a highway maintainable at the public expense.

The great majority of highways in the country are maintainable at the public expense. There are a few highways which are still privately maintainable by reason of tenure, enclosure or prescription and there are a number of highways which are maintainable by no-one. The position with regard to maintenance and repair of the publicly maintainable highways will be considered first; the position with regard to these other classes of highway will then be explained.

**Enforcement of the duty to maintain**

Individuals could not, until 1961, sue the Highway Authority for injuries sustained through failure of the Authority to carry out its duty to maintain the highway. On the other hand, it has been the position at common law for centuries that any person could apply to the Courts for an order compelling the inhabitants at large, and subsequently the Highway Authority, to carry out their duty to maintain the highway. Since the 18th century, and probably long before that, the Royal Justices had a duty to see that the King's subjects could pass freely on the highways. They might be stopped from doing so by obstructions put there or by encroachments on it or by want of repair. Those who in one way or another, or all of these three ways, interfered with the highways were liable to be indicted before the Courts.

So far as want of repair is concerned, section 56 of the Highways Act 1980 now contains the procedure by which any person may apply to the Court for an order requiring the Highway Authority to put a highway in proper repair. The person seeking the order, referred to as the "complainant" in section 56, first serves a notice on the Highway Authority requiring it to state whether it admits that the way or bridge is a highway and that it is liable to maintain it. It should be noted that the section refers throughout to 'the way or bridge', making it clear that bridges are included within its provisions. If within one month from the service of the notice the Authority *does not admit* that the way or bridge is a highway and that it is liable to maintain it, the complainant may apply to the Crown Court for an order requiring the Authority to put the way or bridge in proper repair. The Court first considers whether the way or bridge is a publicly maintainable highway. If it finds that it is, and that it is out of repair, it will make an order for repairs to be carried out within a time specified in the order.

If the Authority serves on the complainant a notice admitting that the way or bridge is a highway and that it is liable to maintain it, the complainant can, within six months of the date of service on him, apply to a local Magistrates' Court for an order requiring the Authority to put the way or bridge into proper repair within a period specified in the order if the Court finds that the highway is out of repair.

If the Authority fails to complete the repairs within the period specified by the Crown Court or the Magistrates' Court, as the case may be, then, unless the Court agrees to extend the period allowed, it *must* authorise the complainant to carry out such works as may be necessary to put the highway into proper repair. Any expense which the complainant then incurs in carrying out the works can be recovered from the Authority as a debt.

This procedure has not been much used during this century when, in general, Highway Authorities have fulfilled their duties to keep highways in repair. Given the limited resources available to Highway Authorities, it is the footpaths and bridleways which are most likely to be out of repair and it is not surprising that the recent reported cases on section 56 concern maintenance of such rights of way. In *Hereford and Worcester County Council v. Newman* heard by the Court of Appeal in 1975, Mr. Newman was seeking to

C

maintain a decision which he had won from the Redditch Magistrates, and which had been upheld by the Divisional Court, that repairs be carried out to certain footpaths to remove a hawthorn hedge and undergrowth growing across paths and a barbed wire fence. The Court upheld the orders in so far as the hedge and undergrowth were concerned but drew an important distinction between the surface of a highway being out of repair and obstructions placed across the highway. They found that a highway gets out of repair because the Authority has not done its duty over a long period and whilst this might be the case with regard to the growth of vegetation in the surface of the paths, it was not necessarily so in the case of an obstruction which could be placed across the highway very quickly. There were many other provisions in the Highways Acts dealing with positive acts of interference by way of obstruction, nuisance and encroachment. It was clear from the existence of these provisions that the 'out of repair' provisions in section 56 were intended to deal with acts of omission on the part of a Highway Authority consisting of neglect and failure to maintain and repair. The highway could only be said to be out of repair if its surface was defective or disturbed in some way. If it had become unusable because of an act of obstruction, it was not out of repair and the removal of the obstruction was not in itself a repair.

Lord Justice Cairns said:

> "I consider that a highway can only be said to be out of repair if the surface of it is defective or disturbed in some way. Not every defect in the surface would constitute being out of repair – e.g. an icy road would not be in my view be out of repair."

The Court of Appeal therefore quashed the order of the Justices with regard to the barbed wire fence but left the remainder of the order in place.

There have been occasional cases where carriage roads which have long since fallen into disuse have been shown to be ancient highways in existence before 1835, often by reference to old maps. In such cases the Highway Authority will be in difficulty in resisting a successful application to the Court that the highway should be put back into a state of repair. If the Authority considers that one individual or company is using this procedure to secure the maintenance at the public expense of a highway which is no longer of any value to members of the public generally, for example, because the village served by the road has disappeared, it can make an application to the local Magistrates under section 47 of the Highways Act 1980 for an order declaring that the highway shall cease to be maintained at the public expense on the grounds that it is "unnecessary for public use". The section does not apply to trunk roads, footpaths or bridleways. If the application by the Highway Authority is successful, the status of the way as a highway open for use by the public remains but the duty on the Highway Authority to maintain the way is removed.

## Damages for failure to maintain

Most disputes about failure to maintain the highway arise from accidents in which users of the highway sustain personal injuries. Liability to pay damages for personal injuries caused through failure to maintain the highway or caused by negligent repair work will be considered in detail in the next chapter.

Cases do occur from time to time when a failure to maintain a highway results in inconvenience and economic loss to owners or occupiers of premises served by the highway. They can, as explained above, seek an order that the Authority should put the highway in repair. It does not appear that damages can be recovered for pecuniary loss for failure by the Highway Authority to maintain a highway, unless the Authority has committed some positive act of obstruction or interference amounting to a nuisance. A Wiltshire farmer succeeded in persuading a High Court Judge in 1991 that he should recover compensation of over £70,000 from the County Council because the highway leading to his farm had fallen into disrepair, and eventually, the Milk Marketing Board had refused to allow their tanker to use it, with the result that he had to give up his dairy business. However, on appeal,[1] the Court of Appeal held that the farmer should have used his statutory remedy of an application

---

1 *Wentworth v. Wiltshire County Council* (1992).

to the Crown Court for an order under section 56 of the 1980 Act that the road should be put into repair. He could not obtain damages for the loss of his business as an alternative remedy.

### Recovery of maintenance expenditure due to extraordinary traffic

At common law, use of a highway by vehicles of excessive weight or use by traffic in unusually heavy volumes by one person could amount to an actionable nuisance. An action could be brought by any individual or by the Highway Authority for the damage done to the highway. Section 59 of the Highways Act 1980 now gives a statutory remedy to the Highway Authority to recover expenses from the person by whose order the extraordinary traffic has been conducted. Subsection 1 provides:

> "Where it appears to the Highway Authority ... that having regard to the average expense of maintaining the highway or other similar highways in the neighbourhood, extraordinary expenses have been or will be incurred by the Authority in maintaining the highway by reason of the damage caused by excessive weight passing along the highway, or other extraordinary traffic thereon, the Highway Authority may recover from any person ('the operator') by or in consequence of whose order the traffic has been conducted the excess expenses."

This statutory power was first enacted in the Highways and Locomotives (Amendment) Act 1878. Most of the cases concerned with the statutory remedy for Highway Authorities occurred around the turn of the century when steam driven locomotives and then motor cars were something of a novelty. With the heavy volume of vehicular traffic using most highways nowadays, it is difficult to show that wear and tear or damage caused to the road has been due to the traffic conducted by any one operator. Moreover, the traffic put upon a road by any one person is not necessarily extraordinary because he uses the road more than others. In order to render him liable, his traffic must be shown to be extraordinary as regards the ordinary uses of the road as a whole

36

by all who use it.[1] Traffic which is extraordinary on any particular road may in the future cease to be so and become ordinary if the road comes to be used by much other traffic. Whilst it may be possible to obtain a contribution towards additional expenses from an operator in one year, it may not be possible to obtain a similar order against him two or three years later if his traffic is no longer extraordinary in relation to other traffic on the road.[2]

The calculation of the 'extraordinary expenses' which the Highway Authority is to recover also leaves a great deal of scope for argument. The sums actually expended by the Authority in previous periods will not be a proper basis of comparison unless they would have been reasonably sufficient for the requirements of the highway in the past. If the highway is so badly damaged that it is necessary to re-make it entirely, the operator can be required to bear the whole expense, but where the Highway Authority has taken the opportunity to improve the highway, the Court will have to apportion the expense as best it can in the event of a dispute.

Proceedings to recover extraordinary expenses under section 59 are taken in the County Court and have to be brought within twelve months from the time when the damage was done or, if the damage is due to a particular building contract extending over a period, then not later than six months from the date of completion of the contract. If use of the road by extraordinary traffic can be foreseen and if the operator can be persuaded to admit liability then, under subsection 3 of section 59, there is a useful power for the Highway Authority and the operator to agree the payment of a sum by the operator by way of a composition of his liability under the section. Alternatively, so long as the operator admits liability, either he or the Authority may require the sum to be paid to be determined by arbitration.

## Highway physically destroyed

There have been no reported cases this century on whether a Highway Authority has a duty to rebuild a road which has been destroyed, for example, by the sea. In the old cases, it was held that

---

1 *Hill v. Thomas* 1893.

2 *Henry Butt & Co. Ltd. v. Weston-super-Mare U.D.C.* 1922.

where the actual site of the road, whether on natural ground or on artificial embankment, had been destroyed, the liability to maintain was extinguished.[1] If, however, a landslip has destroyed the surface but not the site or foundation of a road, it is a question of fact whether the road could be said to have been destroyed.[2] It also appears that in the past the Courts have taken a practical view and have taken into account the cost of reinstatement in proportion to the value of the road. If a highway runs along a seashore, an embankment, sea wall and groynes may be necessary for its protection and, if so, will have to be repaired by the Highway Authority as part of their obligation to maintain the highway.[3]

Where a highway is supported by retaining walls, as for example along the side of a hill, those walls may be part of the highway and repairable as such.[4]

### Privately maintainable highways

A person or corporation can be liable to maintain a highway by reason of statute, tenure, enclosure or prescription. An obligation to repair a highway may have been imposed by statute. Liability by prescription arises where for a number of years it can be shown that private persons and their predecessors have repaired the road in question. Proof of some consideration, such as the right to take tolls, is usually required in order to establish this liability. Liability by reason of tenure arises out of the ownership of adjoining land to which a duty to repair the highway has been attached. It is usually sufficient to demonstrate that for a number of years the owners concerned and their predecessors, or their tenants, have repaired the road in question. Sometimes a liability to repair by reason of ownership carried with it an exemption for the owners and occupiers from liability in respect of rates and similar local taxes for the repair of other highways in the parish. Liability by reason of enclosure arises where a highway crossed unenclosed land and the

---

1  *R. v. Paul Inhabitants* 1840.

2  *R. v. Greenhow Inhabitants* 1876.

3  *Sandgate U.D.C. v. Kent County Council* 1898.

4  *R. v. Inhabitants of Lordsmere* 1886.

public acquired a right to deviate onto that land when the highway was impassable. If then the owner enclosed his land by a fence, he or his tenant became liable by reason of the enclosure to keep in repair the original line of the highway and make it passable at all times.

Where a person is liable by reason of tenure, enclosure or prescription to maintain a highway, either he or the Highway Authority can apply to the local Magistrates for an order under section 53 of the 1980 Act that the liability of that person to maintain the highway shall be extinguished. If the order is made, the highway then becomes a highway maintainable at the public expense. The Highway Authority have the right to be heard by the Magistrates where the application is made by the person responsible for maintaining the highway.

Where the Magistrates make an order extinguishing the private liability to maintain the highway, the person whose liability is so extinguished has to pay to the Highway Authority a sum to be agreed between them, being either a lump sum or annual payments for a number of years. If agreement cannot be reached between the parties, the matter is referred to an arbitrator appointed by the Secretary of State.

## Powers and duties of persons liable to maintain highways

Persons liable to maintain highways privately have many of the powers and duties of the Highway Authority but there is no general provision in the Highways Acts which says this. Instead, individual sections in the Acts will say that their provisions are applicable to privately maintainable highways as well as to publicly maintainable highways. Thus, identical powers to obtain materials to repair highways from waste or common land exist for both Highway Authorities and those liable to maintain highways privately. They have the same powers to construct surface water drains in the highway or in land adjoining the highway. They have the same duties to remove snow, or soil from landslips, from the surface of the highway.

As is mentioned earlier in this chapter, any person can apply to the

Courts for an order compelling a Highway Authority to put into repair a highway which is maintainable at the public expense. This power arises under section 56 of the Highways Act 1980. The following section, section 57, provides that with regard to highways which a person is liable to maintain privately, the Highway Authority can step in and carry out repairs if, in their opinion, the highway is not in proper repair. They can then recover the necessary expenses of so doing from the person responsible for the maintenance of the highway. This power cannot be exercised until the Highway Authority have given notice to the person liable to maintain the highway that it is not in proper repair and have allowed him a reasonable time within which he can repair it.

It is still an open question whether individuals or bodies responsible for privately maintainable highways can be sued by a person who has sustained injury through a failure to repair the highway.[1]

### Highways repairable by no-one

Privately maintainable highways are not found very frequently in practice. Also, where a person is clearly responsible for the maintenance of a highway, he has a strong interest in removing that liability by coming to an agreement with the Highway Authority for his liability to be extinguished under the section 53 procedure described above. The highway engineer or legal practitioner is more likely to come across highways maintainable by no-one. Many roads have been constructed since 1835 to a poor standard which the Highway Authority has never agreed to adopt as publicly maintainable. Nowadays, purchasers of houses and their building societies are likely to look askance at properties served by roads which are not maintainable at the public expense. Builders, therefore, routinely seek to enter into an agreement with the Highway Authority under section 38 of the Highways Act 1980, for the construction of the road to an acceptable agreed standard followed by its dedication and adoption by the Highway Authority. This position has only been the norm since the enactment of the Highways Act 1959.

---

1 *Rundle v. Hearle* 1898.

There are quite a number of housing estates laid out before 1959 where the roads are maintainable by no-one. The roads may not have been formally dedicated as highways by the builder who originally owned all the land and laid out the estate. In practice, if there has been no attempt by the builder or the residents to stop the general public using the roads over a period of twenty years, they will be deemed to have been dedicated as highways under section 31 of the 1980 Act, as explained in Chapter 2. The Highway Authority has no right to execute repairs in such roads except under certain specific statutory powers. The section 57 power to step in and put the road in repair applies only when the Authority can show that someone is responsible for the maintenance of the road by statute, tenure or prescription. This is not usually the case with housing estates laid out during this century.

Under section 230 of the Highways Act 1980, where repairs are needed to 'obviate danger to traffic' the Authority may by notice require the owners of the premises fronting the street[1] to execute repairs whether or not the street is a highway. Such repairs tend to be of a limited nature such as patching or filling potholes. The frontagers can appeal to the local Magistrates against such requirements. If the requirements are not appealed against, or are upheld on appeal, the Authority can execute the repairs and recover the costs from the owners in default. Alternatively, the majority of the owners of premises in the road (the frontagers) can require the Authority to make up the street at the expense of the frontagers under the Private Street Works Code described in Chapter 8 and thereafter maintain it at the public expense.

Under the Private Street Works Code the expenses of making up a private street fall upon the owners of the properties fronting the street – the frontagers. These expenses can be quite heavy. It has been held that the frontagers, even if the builder has retained the ownership of the road, have an interest in preventing damage to the private street. In one case they were able to obtain an injunction forbidding a company owning a horse-drawn baker's van from

---

1  Section 230 uses the word 'street' and it has, therefore, been used here. 'Street' is defined in section 329 of the 1980 Act and includes any highway and any road. See also the explanation of various definitions in Chapter 1.

attaching a skid-pan to the wheels of the van in order to stop the horses trotting away with the van. The skid-pan was causing deep ruts in the road and was making the road more difficult for the frontagers to use. It was also making it more likely that the Highway Authority would bring forward the date for carrying out private street works at the expense of the frontagers.[1] Hence they had an interest in preventing damage to the road.

The ownership of highways will be discussed in Chapter 5. In the case of highways repairable by no-one, only the owner of the soil of the highway is entitled to carry out repairs or improvements. There may be an identified owner, who can show title to the whole length of the road. Alternatively, the frontagers may each own half of the road outside their frontage, up to the middle line of the road. Apart from the emergency works referred to above under section 230 of the 1980 Act, or from carrying out a full scheme under the Private Street Works Code, the Highway Authority has no right to carry out repairs or improvements in a highway which is not publicly maintainable.[2] Residents or frontagers who want to improve parts of the road which are not outside their own properties, for example by resurfacing or metalling, will be unable to do so without the permission of the owner of those lengths of the highway, unless their title deeds contain private rights of way for them to use the whole road. In that event, the Courts will imply a right to carry out repairs and improvements, so that the private right of way can be exercised.[3] This, however, would not be a matter of highway law, but of private rights as between neighbouring landowners.

---

1  *Medcalf v. R. Strawbridge Ltd.* 1937.

2  *Campbell Davys v. Lloyd* 1901.

3  *Newcomen v. Coulson* 1878.

# Chapter 4

# Liability of the Highway Authority for Accidents on the Highway

## The standard of maintenance expected

In the last chapter the duty on the Highway Authority to maintain the highway was discussed and the statutory remedies for non-repair were explained. Highways Acts have not stipulated the standard of maintenance which should be applied to highways and, indeed, it is accepted by the Courts that different standards will apply to major roads, minor roads, footpaths and bridleways. It is necessary to look at a substantial body of case law to get a 'feel' for the standards which the Courts will apply. These cases have usually arisen as a result of accidents sustained by a plaintiff who has then brought an action against the Highway Authority for breach of its statutory duty to maintain the highway. Section 41 of the Highways Act 1980 simply says that the Highway Authority are under a duty 'to maintain the highway'. The interpretation section in the Act, section 329, provides that 'maintenance' includes repair and the word 'maintain' is to be construed accordingly.

## Degree of care to be taken by drivers

Highway Authorities cannot expect all drivers to show a high degree of skill in driving their vehicles along the highway, although in quite a few cases Courts have reduced damages awarded against Highway Authorities because of the contributory negligence of a driver in not paying due regard to road conditions. In the case of *Rider v. Rider* heard by the Court of Appeal in 1972, Lord Justice Sachs said:

"The Corporation's statutory duty under section 44 of

the 1959 Act (the predecessor of section 41 of the 1980 Act) is reasonably to maintain and repair the highway so that it is free of danger to all users who use that highway in the way normally to be expected of them – taking account, of course, of the traffic reasonably to be expected on the particular highway. Motorists who thus use the highway, and to whom a duty is owed, are not to be expected by the Authority all to be model drivers ... The Highway Authority must provide not merely for model drivers, but for the normal run of drivers to be found on their highways, and that includes those who make the mistakes which experience and common-sense teaches are likely to occur. In these days, when the number and speed of vehicles on the roads is continually mounting and the potential results of accidents due to disrepair are increasingly serious, any other rule would become more and more contrary to the public interest."

### What must a road user prove to succeed in a claim for damages?

*The 'special defence' for Highway Authorities*
Lord Denning said in *Burnside v. Emerson*, heard in 1968, that an action for damages for injuries caused through breach of statutory duty to maintain the highway, involves three things. First, the plaintiff must show that the road was in such a condition as to be dangerous for traffic. Second, he must prove that the dangerous condition was due to the failure by the Highway Authority to maintain, which would include a failure to repair the highway. If the plaintiff succeeds on these two issues, the Authority is at first sight liable for any damage resulting therefrom. However, it has a special defence in an action for failure to maintain contained in section 58 of the Highways Act 1980. If it can prove that it had taken "such care as in all the circumstances was reasonably required to secure that the part of the highway to which the action relates was not dangerous for traffic" the plaintiff will fail. Before 1961 Authorities were only liable for injuries caused through carrying out their duties negligently. They were *not* liable to individual users

of the highway for injuries caused through failure to maintain or repair the highway, that is to say for their acts of omission or inactivity. When Parliament abrogated this old common law rule in enacting the Highways (Miscellaneous Provisions) Act 1961, they afforded this special defence of having taken such care as in all the circumstances was reasonably required so that failure to maintain would not in itself necessarily involve an Authority in liability for accidents. It should be noted that this defence only applies in respect of the failure to maintain or repair and would not be available to an Authority in a claim in respect of an accident arising out of repairs negligently carried out by the Authority.

It should also be noted that, in seeking to establish this special defence, an Authority cannot hide behind the acts of a contractor appointed by it to carry out maintenance or repair work unless they can also prove that they have given him 'proper instructions' and 'that he had carried out the instructions'.

In considering whether the special defence is applicable, the Court is to have regard to the following matters specified in subsection (2) of section 58:

> The character of the highway and of the traffic expected to use it.

> The standard of maintenance appropriate for a highway of that character.

> The state of repair in which a reasonable person would have expected to find the highway.

> Whether the Highway Authority knew, or could reasonably have been expected to know, that the condition of the part of the highway to which the action relates was likely to cause danger to users of the highway.

> If the Highway Authority were aware of the danger and could not have been reasonably expected to repair that part of the highway before the accident, what warning notices of the dangerous conditions had been displayed.

45

## Cases on the breach of the statutory duty to maintain

The following cases illustrate the way Courts have applied these principles:

> *Griffiths v. Liverpool Corporation (1966)*
> The plaintiff had tripped over a paving stone protruding half-an-inch above other paving stones. It was a loose paving stone which rocked on its centre and it was agreed by the surveyors on both sides that the paving stone was dangerous. The Corporation pleaded the special defence but the County Court Judge, whose decision was upheld by the Court of Appeal, found that reasonable care had not been exercised. Had there been a regular system of inspection every three months, the dangerous paving stone would have been noticed and would have been made safe.

> *Meggs v. Liverpool Corporation (1968)*
> In this case the plaintiff tripped whilst walking along a pavement with uneven paving stones. They had sunk in different places, so much so that one of them had sunk about three-quarters of an inch. This time the Highway Authority succeeded in establishing the special defence of reasonable care. The Judge said, "On the totality of evidence and looking at the reality of the situation that thousands use the footwalk, that no-one has avoided it or reported it, I am not satisfied that there was such a danger present as to indicate a failure on the part of the defendants to discharge their duty." The Court of Appeal said that this was a finding of fact on the part of the County Court Judge and refused to reverse it.

> *Littler v. Liverpool Corporation (1968)*
> Again, this was a case of the plaintiff tripping over the lip of a paving stone sticking up above the level of the pavement along which he was coming. He was running at the time. The Judge said:

> "The test in relation to a length of pavement is reasonable foreseeability of danger ... it is a mistake to

isolate and emphasize a particular difference in levels between flagstones unless that difference is such that a reasonable person who noticed it and considered it would regard it as presenting a real source of danger ... a highway is not to be criticised by the standards of a bowling green."

The plaintiff's claim was dismissed.[1]

*Bird v. Tower Hamlets London Borough Council (1969)*
The plaintiff, who delivered newspapers to shops and stalls, parked his motorvan outside a shop, tried to alight and put his left foot into a hole in the road approximately twelve inches by six inches wide and three inches deep, injuring his ankle. The Judge observed that when that particular road was re-surfaced, the machine used to do it took three lanes in order to cover the surface and between the lanes there were joints. He was satisfied that the joints were a vulnerable point and that potholes were likely to appear there. The Authority's own engineer said that, while a hole of such a relatively small depth as three inches might not be unsafe for vehicles, it could constitute some slight danger to pedestrians. He conceded that if he had been aware of the hole, he would "at once get it filled in for that reason". The Judge said that his personal inclination was to find that a hole of this shallow depth might more appropriately be called a depression but he felt that in the light of the evidence, he must find a condition of danger and that the defendant Council was liable.

*Ford v. Liverpool Corporation (1972)*
In this case the plaintiff was injured when she tripped over the surround of a metal grid in the carriageway. The grid was five feet from the pavement and 43 inches long and eighteen inches wide. At the edge of the grid, furthest from the pavement, the tarmac stood proud of

1  In *Mills v. Barnsley M.B.C.* 1992 the Court of Appeal has again stated that the public must expect to find minor defects in pavements and must look where they are walking.

the grid by just over one inch. The Judge said that the actual depth of the obstruction was not the only criterion; other matters such its position, the kind of roadway in which it lay, the use made of the roadway and the obvious nature of the obstruction had to be taken into account. Whilst the plaintiff argued that the ridge was more dangerous in a roadway than in a pavement, the Judge did not agree. It seemed to him to be more important that the pavement should be free of obstructions. The carriageway had on it a number of things which could be called obstructions, such as cobblestones, cat's eyes and pedestrian crossing studs. The general public must expect obstructions to appear in the roadway and such obstructions were one of the normal hazards of life. He found in favour of the defendant Council and dismissed the claim.

*Burnside v. Emerson (1968)*
This was a case involving an accident between two motor vehicles driving along a main road in very wet weather. The defendant was driving his Rover car and ran into a pool of water which was half-way across the road. As a result he went right across the road into the path of the plaintiff's oncoming Jaguar. Mr. Emerson, the driver of the Rover, was killed. Mr. Burnside, the plaintiff, the driver of the Jaguar, and his wife, suffered very serious injuries and brought an action against the Executors of Mr. Emerson's estate. They also joined the Nottinghamshire County Council, who were the Highway Authority, as the second defendants because they alleged that they had not done their duty in regard to the highway in that they had not drained the road properly. The Judge, at first instance, considered that the County Council was wholly liable. On appeal, Lord Denning said, "when there is a transient danger due to the elements, be it snow or ice or heavy rain, the existence of danger for a short time is no evidence of a failure to maintain". However, evidence had been called which showed that this stretch of road was not kept

properly drained. It was quite often flooded when there was rain. The Court went on to consider whether the Authority had established the special defence and decided that, as the County Council could have been expected to know that the road always flooded after rain, they had not discharged the burden of proving that they had taken all such care as was reasonably required.

The Court of Appeal did, however, reverse the lower Court's finding that the first defendant, Mr. Emerson, the driver of the Rover, was in no way to blame. They considered that the fault was two-thirds on his part and one-third on the part of the Highway Authority and allowed the appeal to that extent.

*Tarrant v. Rowlands (1978)*
This was a similar case on flooding to that of *Burnside v. Emerson* except that in this case the collision between two vehicles had occurred on a trunk road, the A6 in Hertfordshire, and the Department of Transport were joined in the action as the Highway Authority. The Judge accepted the principles laid down in *Burnside v. Emerson*. On the facts of the case, the Judge accepted evidence that there was a pool of water at the point of the accident after heavy rainfall. Although the Highway Authority claimed to be unaware of it, he found that they could have been aware of it, had they carried out regular inspections and that the special defence failed because they could have been expected to know that the condition of that part of the road was likely to cause danger to traffic. In this case he found that Mr. Rowlands had been driving slightly more slowly than Mr. Emerson and that the A6, being the ancient highway between London and Carlisle, was a more important road than the road concerned in *Burnside v. Emerson*. Accordingly, he apportioned the liability between the Highway Authority and Mr. Rowlands as half-and-half, rather than two-thirds against the driver, as in *Burnside v. Emerson*.

*Haydon v. Kent County Council (1977)*

Mrs. Haydon slipped on ice on a steep public footpath in February 1973 during a period of snowy weather. The public footpath was tarmacadamed and was much used. She brought an action against Kent County Council for damages for her injuries alleging failure on the part of the County Council to fulfil its statutory duty to maintain. The majority of the Court of Appeal decided that the statutory obligation to maintain "does include clearing snow and ice, or providing temporary protection by gritting, but whether there has been a breach of this duty is a question of fact and degree on the facts of each particular case". The Court had regard to the fact that the snow had fallen three days previously and, in the words of Lord Justice Goff, "one must not lose sight of the heavy commitments that the Highway Authority had to keep major highways and other important roads safe and clear". Moreover, the attention of the Authority had not been drawn to the dangerous conditions on the path until the morning of the accident. Lord Justice Goff, and Lord Justice Shaw who agreed with him, both decided that in the circumstances the Local Authority had not failed in their duty to maintain the footpath and that there was no need to consider the special defence afforded by section 58. Indeed the County Council had deliberately not pleaded this defence but had rested their case on the argument, which the Court upheld, that there had not been a failure to maintain.

Lord Denning delivered a dissenting judgement in which he argued that the Highways Act 1959 (and the same applies to the later Highways Act 1980) was a consolidating statute and that prior to 1959 the duty of the inhabitants at large, and subsequently the Highway Authorities, was to repair the highway. Therefore, in his view, 'maintain' in the 1959 Act could not mean more than repair. Clearing away snow was not a repair and therefore the Authority was not under any duty to do it.

The majority of the Court disagreed with this view. Whilst all three Judges agreed that the presence of snow and ice on the highway did not indicate a lack of repair and that clearance of snow and ice was a matter of maintenance, the majority of the Court held that the statutory obligation to maintain did include clearing snow and ice. If the highway in question had been a major road and had not been cleared and gritted within three days, the majority would almost certainly have found the Authority to be in breach of its statutory duty to maintain.

*Bartlett v. Department of Transport (1984)*
This case again concerned an accident in winter when there was a collision between cars travelling on the A34 trunk road between Newbury and Oxford. The trunk road had not been gritted even though the icy conditions had been in existence for up to three weeks. This was because of an industrial dispute. The Trade Union concerned had said that, in furtherance of a pay dispute, their men would not salt and grit the A34 but would carry out this work on all other roads in Oxfordshire. If, however, the Highway Authority brought in independent contractors to work on the A34, the Union would instruct their men not to carry out maintenance work on the other roads.

The Judge followed the majority in *Haydon v. Kent County Council* in deciding that there was a duty to grit and salt the A34 road. He decided that the Department of Transport were not in breach of this statutory duty because they owed a duty to all road-users and they would have made the position worse if they had used independent contractors on the A34. The Judge went on to find that, if he had decided that the Authority were in breach of their statutory duty to maintain, they could have successfully pleaded the statutory defence of reasonable care because they had displayed very large signs on the roadside saying, "Road not gritted". They

51

had also sought assistance from neighbouring
Authorities, the police and motoring organisations to
direct as much traffic as possible away from the A34 and
had given warnings of conditions on the radio. Finally,
the Judge found that the driver had been overtaking in
these treacherous conditions and going too fast. Had the
driver's widow succeeded in establishing liability
against the Department of Transport as the Highway
Authority, he would have reduced the damages by 80%
due to the substantial contributory negligence of the
driver.

### Rider v. Rider (1972)

Mrs. Rider was injured when the car in which she was
being driven by her husband along a country lane
swerved and collided with another vehicle. It was agreed
before the trial that Mrs. Rider was entitled to recover
against either her husband or the Southampton
Corporation as the Highway Authority and the real issue
lay as to liability between those two defendants. A
particularly interesting feature of the case was that the
road concerned was a narrow country lane winding
between villages. As time went on motor traffic had
increased generally and it had come to be used as an
escape route or 'rat run' by drivers who found the A27
road between Romsey and Southampton jammed with
traffic. As a result it had become regularly used by
vehicles of all descriptions, including lorries. All this
was known to the Highway Authority but they did
nothing to keep the road in repair or improve it. The
edges led onto grass or mud verges which were in places
below the level of the tarmac. As it was necessary for
vehicles passing each other to hug the unsupported
edges, chunks were broken off. No steps had been taken
to bring the lane up to the standard required for a
secondary main route. Indeed, for some time its
condition had been allowed to deteriorate, probably
because a decision had been taken by the Authority to
build another road to take through traffic from Romsey

to Southampton. This situation will be all too familiar to highway engineers who have to cope with increasing volumes of traffic, restrictions on public expenditure and the propensity of drivers to take to the most minor roads if they can avoid traffic jams.

The Highway Authority contended that Mr. Rider's car had swerved across the road because he had over-corrected a tail skid caused by driving too fast. It was contended for Mr. Rider that the swerve was due to one of his nearside wheels going into an indentation in the edge of the lane and that the swerve was, therefore, caused by the state of disrepair. The Court accepted this argument and for that reason had to consider whether the state of disrepair was such as to constitute a breach of the Authority's duty. The Highway Authority argued that a sufficiently careful driver would not have been put at risk by the state of the lane and would have taken its condition into account. The Court rejected this argument saying that it is the duty of the Authority to maintain and repair the highway so that it is free of danger to all users, taking account of the traffic reasonably to be expected on the particular highway. Motorists are not to be expected by the Authority to be model drivers. In the words of Lord Justice Sachs, quoted earlier in this chapter, "in these days, when the number and speed of vehicles on the roads is continually mounting and the potential results of accidents due to disrepair are increasingly serious, any other rule would become more and more contrary to the public interest." In this case there was no direct evidence of negligence. The trial Judge had, however, decided that Mr. Rider was one-third responsible for the accident, probably because of his local knowledge of the condition of the lane. The Court of Appeal seemed more inclined to find the Highway Authority entirely liable but decided not to interfere with the trial Judge's apportionment.

This case indicates that the statutory duty to repair is not confined to keeping the road in the state of repair in which it was in, say, 1835 or 1935. If it is a highway which has been dedicated for use by wheeled traffic and traffic has greatly increased over the years, then the Authority will have a duty to strengthen the highway. This may include going so far as to remake or repair with stone a road which in the past has never had a hard foundation. Indeed, in a case heard as long ago as 1892,[1] the Court of Appeal held that a landowner over whose land a highway had been dedicated by himself or his predecessors in title, could not object to the Highway Authority metalling the road which had previously been an unsurfaced track.

### Importance of factual evidence at trial

In practice, each case turns on its own facts. Impressions which the plaintiff and the defendant Highway Authority's officers and workmen make on the trial Judge hearing the case are often all important. The Court of Appeal does not hear witnesses and will not overturn a finding of fact by a trial Judge: for example, the finding in *Burnside v. Emerson* that the drain had not been put in at the lowest point in the road and the finding in *Rider v. Rider* that the driver swerved because one of his nearside wheels went into an indentation in the edge of the lane.

### Conclusions to be drawn from these cases

The decisions in Haydon's case and Bartlett's case established that the section 41 duty to maintain the highway is not an absolute duty. The commitments of the Highway Authority in having to maintain a vast mileage of roads and footpaths, and circumstances beyond its control, such as industrial action, can lead to a successful defence against an action for failure to maintain without the Authorities having to rely on the special defence in section 58. If the Authority has to fall back on that defence, it will be unsuccessful if it could have foreseen the possibly dangerous state of affairs, as the flooding

---

1 *Eyre v. New Forest Highway Board* 1892.

cases show. It will also be unsuccessful, as Rider's case shows, if the standard of maintenance to which the Authority has worked is inappropriate to a road taking account of its importance in the highway network and the volume of traffic using it. Finally, while the Highway Authority, or in practice their insurers, may successfully plead contributory negligence on the part of the driver or drivers concerned in an accident, they cannot necessarily escape all liability by arguing that the drivers should have noticed that the road was dangerous and should have taken more than ordinary care for that reason.

## Negligence: poor performance of work in building or repairing the highway

All the cases discussed above depend upon whether the Highway Authority had carried out its duty under section 41 of the Highways Act 1980 to maintain the highway. They turn on alleged acts of omission, for example where the Authority failed to repair pavements or maintain a proper system of drainage. Users of the highway have always been able to bring an action in negligence against the Highway Authority if it has carried out its duty to maintain the highway in a negligent way as a result of which a person has suffered injuries. In order to win an action for negligence the plaintiff must show that there is a relationship between the plaintiff and the defendant which requires the defendant to exercise due care to avoid the plaintiff being injured. The Courts have held that there is such a relationship between users of the highway and the Highway Authority.

In the case of *Levine v. Morris* (1969), the Department of Transport was held to be liable for the negligent siting of a large road sign close to a bend on a fast dual carriageway. Mr. Morris was driving a car in which Mr. Levine was a passenger. Mr. Morris had lost control of the car in heavy rain and collided with the concrete columns of the sign. Mr. Levine was killed. Again, the Judges stressed that all motorists can be guilty of errors of one kind or another and that accidents were likely to happen close to bends, sometimes of course through no fault at all, such as a burst tyre. The Court of Appeal held that the siting of the sign four feet off the end

55

of a bend was foreseeably likely to give rise to a hazard. They held that in that situation, the Highway Authority had a duty to motorists to consider alternative sites and, where there were two sites equally good as regards visibility, not to select the one that involved materially greater hazards to motorists. The Court held the driver was also liable for driving fast in heavy rain and skidding out of control when he tried to pull in after carrying out an overtaking manoeuvre. The driver was held to be liable as to 75%. Payment of damages was apportioned between the Department and Mr. Morris.[1]

In the case of *Lewys v. Burnett* (1945) the defendant Highway Authority had reduced the headroom under a low bridge to eight feet nine inches by respraying the road and raising its level by six inches. They were held liable to pay damages in an action by the widow of a soldier. He had been standing in a lorry in which he had been offered a lift and had struck his head on the bridge and died. Although there were warning notices, the Court held that these would not necessarily be visible to a vehicle travelling at speed.

In *Bright v. Ministry of Transport* (1970), the Ministry had removed continuous double white lines separating two lanes of a road going uphill from one lane going downhill in the opposite direction. They had replaced the double white lines by a broken line in the centre of the road. The work was done by covering the white lines and the gap between them with cold asphalt. No instructions were given that special measures should be taken to compact the area between the lines so as to secure a continuing surface free from grooving. In the result, there were two long longitudinal ridges along the road, each about half-an-inch proud of the road surface. Mr. Bright was a young man, aged eighteen, on a motor-cycle. The ridges caused his motor-cycle to wobble and go out of control and he suffered serious injuries. The Judge found that the ridges were a potential source of danger and that the likelihood of accident could have been foreseen by the Ministry. He decided that the Ministry were liable in negligence and awarded damages against the Minister to Mr. Bright.

In litigation on accidents on the highway, lawyers acting for drivers

---

1  See also *Bird v. Pearce* 1978 discussed in Chapter 6.

will often find it useful to seek out the relevant technical memoranda or guidance notes issued by the Department of Transport on the design of highways. These contain much detail on all aspects of highway design such as, for example, the extent of visibility required on bends and junctions and the height and location of crash barriers. If a road has not been designed or maintained according to these recommendations, it may be possible to establish that some part of the responsibility for the accident should fall on the Highway Authority.

## Nuisance

The creation of an obstruction or danger on the highway is a public nuisance. If a person can show special damage because, for example, he suffered personal injuries as a result of the danger he can succeed in an action for damages for nuisance. Trees failing across the highway,[1] an unlighted heap of soil on the highway[2] or a pile of slates[3] left at the side of the carriageway over which a pedestrian tripped have all been held to be nuisances. It is necessary in cases of nuisance to show that the defendant was aware of the nuisance and that the nuisance caused the injury but it may not be necessary to show negligence.

In *Dymond v. Pearce* (1972) a lorry driver left his vehicle overnight in a dual carriageway in a built up area. He switched on the rear lights of the lorry and parked it under a street lamp. A motor-cyclist crashed into the lorry during the night. The pillion passenger sued the lorry driver and the owner of the lorry in nuisance. All three Judges in the Court of Appeal held that the parking of the lorry all night in the highway was an actionable nuisance but found that the accident was not caused by the nuisance. They accepted the finding of the trial Judge that the accident occurred because the driver of the motor-cycle "was watching the attractive young ladies on the pavement instead of looking ahead of him to see what conditions he was about to encounter" in the road. The Court left open the

---

1  See Chapter 6.

2  *Penny v. Wimbledon U.D.C.* 1898.

3  *Almeroth v. Chivers* 1948.

question of whether the outcome would have been different if the tail lights had failed or mist or fog had occurred during the night and concealed the lorry.

In an earlier case in 1956 Lord Denning said:[1]

> "As all lawyers know, the tort of public nuisance is a curious mixture. It covers a multitude of sins. We are concerned today with only one of them, namely a danger in or adjoining a highway."

## Choice of cause of action

Thus a person wishing to begin an action for an accident on the highway may be able to found his claim on breach of statutory duty by the Highway Authority or on negligence or nuisance on the part of the Highway Authority or other persons. In an action against a Highway Authority, which had planted a tree in the highway verge and had allowed the branches to overhang the carriageway,[2] the Master of the Rolls said in the Court of Appeal:

> "It does not appear to me to matter very much whether the action is regarded as an action based on breach of statutory duty or merely on nuisance or negligence at common law, because it is common ground that, if provisions of the statute governing the conduct of the Local Authority, on their true interpretation, make it a breach of that statute for the authority to allow branches to grow out in the way these branches grew out, the Authority must be regarded as being in breach of its obligations to the plaintiff."

Breach of statutory duty can be pleaded in those cases where the Authority is alleged to have failed to carry out its duty to maintain the highway. Most claims by drivers against other drivers for bad driving causing accidents are founded in negligence. Negligence can also be pleaded against the Highway Authority as will be seen from some of the examples given above. Nuisance may be the

---

1  *Morton v. Wheeler* 1956.

2  *Hale v. Hants & Dorset Motor Services* 1947.

appropriate cause of action where the Highway Authority or any other person has caused an obstruction on the highway or created the danger.

Actions in nuisance and negligence have always been available against the Highway Authority. An action for breach of statutory duty to maintain the highway has only been possible since 1961 when Parliament repealed the old Common Law Rule that the Highway Authority could not be sued for nonfeasance, the technical term for failure to perform a statutory duty. The special defence that the Authority had taken such care as in normal circumstances was reasonably required was introduced at the same time. It is certain that it is available to the Highway Authority in resisting actions for breach of statutory duty. Text book writers on torts also consider that the Section 58 Defence is available to the Highway Authority in resisting actions for negligence and nuisance though there is, as yet, no authority to support this view. In the case of negligence this is not of much significance since the plaintiff will have in any event to show a failure to take care in order to succeed in a claim of negligence. The point could be important in an action for nuisance where a plaintiff can succeed even though a defendant has not been negligent. There is no decided case as to whether the Highway Authority could plead the Section 58 Defence that it had taken the appropriate degree of care in order to resist successfully a claim in nuisance.

Chapter 5

# Ownership of Highways and Public Rights of Passage

Where the Highway Authority has acquired by agreement or compulsory purchase the freehold of the site of a road, ownership belongs to the Highway Authority. This will usually be the case with motorways and by-passes built in recent years where the Highway Authority has paid cash to buy the land.

The great majority of highways have been created by dedication, usually many years ago. *Ownership of the soil beneath the highway remains with the land owner who originally dedicated it and his successors in title.* The public right of passage is paramount and the owner must do nothing to obstruct or interfere with it, but he retains a number of important rights. The Highway Authority has been given a wide range of powers by Parliament to enable it to maintain the highway, but it must always be careful to act within those powers. If it exceeds them remedies are available in the Courts, both to the owner, and to members of the public.

Neither Parliament nor the Courts have found it easy to define precisely the boundaries of these three overlapping interests. Section 263 of the Highways Act 1980 provides that "every highway maintainable at the public expense, together with the materials and scrapings of it, vests in the Authority who are for the time being the Highway Authority for the highway". Section 335 specifically reserves the rights of owners to the mines and minerals under the highway as if the highway were not vested in a Highway Authority.

The effect of vesting the highways in the Highway Authorities is not to transfer the ownership in the land but to vest in the Authority the property in the surface of the road and in so much of the actual soil below, and air above, as may be required for its control, protection and maintenance as a highway for the use of the public.

## Meaning of 'vest'

The use of the word 'vest' in section 263 of the 1980 Act, quoted above, is not new and it appeared in nineteenth century statutes dealing with highways, such as the Public Health Act 1875. It is an expression which has given Judges difficulty in both this century and the last. Lord Justice Bramwell said in a case heard in 1878:[1]

> "I confess I have my misgiving as to the meaning of the word 'vest' – speaking then with all respect of section 149 of the Public Health Act 1875, which section is nearly a repetition of section 68 of the Public Health Act 1848, I find it very difficult to put a meaning upon that word."

He summed up his view on the meaning of the word by saying:

> "That would show that 'street' comprehends what we may call the surface, that is to say not a surface bit of no reasonable thickness, but a surface of such a thickness as the local board may require for the purpose of doing to the street that which is necessary for it as a street, and also of doing those things which commonly are done in or under the streets; and to that extent they had a property in it."

In a case heard in 1954[2] the phrase 'vest' was still giving difficulty to the Court of Appeal. Lord Denning said:

> "The Statute of 1929[3] vested in the Local Authority the top spit, or perhaps I should say the top two spits, of the road for a legal estate in fee simple determinable in the event of its ceasing to be a public highway."

Thus, following this reasoning, Highway Authorities were not permitted to carry out works below the surface of the road which had nothing to do with their duty to maintain the highway. In a case heard in 1896,[4] which went as far as the House of Lords, it was held

---

1  *Coverdale v. Charlton* 1878.

2  *Tithe Redemption Commission v. Runcorn U.D.C.* 1954.

3  Local Government Act 1929.

4  *Tunbridge Wells Corporation v. Baird* 1896.

that the Highway Authority had no power to excavate land under the highway in order to erect lavatories below the surface of a street which had vested in it within the meaning of the Public Health Act 1875. Authorities could not provide public conveniences on highway land until specifically authorised to do so by the Public Health Act 1936.

There has never been any doubt in the mind of the Judges that Parliament has used the word 'vest' in the statutes to make it clear that full ownership is not being conferred on the Highway Authorities. As the Master of the Rolls put it in the 1954 case:

> "So long as the land is used as a highway, it remains vested in the local authority. On it ceasing to be a highway, the Authority's interest is automatically 'divested' in favour of the original owner or his successors."

In *Wiltshire County Council v. Frazer* (1983) the Court of Appeal, following this case, decided that the Highway Authority had a sufficient interest in the ownership of the highway to be able to bring an action in trespass for an order to recover possession of the highway from squatters who were living in tents and caravans on the highway.

### Supporting structures – ownership

The extent of the Authority's property in structures supporting the highway has given even more difficulty. Lord Denning's use of the phrase "the top two spits" works well enough on a flat road. It does not resolve the extent of the Highway Authority's interest in bridges, embankments, abutments and retaining walls. The Judges have tended to suggest that Authorities have vested in them as much of the subsoil or retaining walls or embankments as is necessary to maintain the highway.

When the High Court had to consider a case in 1933[1] concerning a highway built on a raised viaduct, resting upon open arches and descending to the ground level upon embankments, the Court held

---

1  *Hertfordshire C.C. v. Lea Sand Ltd.* 1933.

that the vesting of the road included the whole masonry and structure of the viaduct together with such soil as it actually occupied and the soil of the embankments but not the soil lying beneath and between the arches of the viaduct. Similarly, if a retaining wall is necessary to support the highway, it would be regarded as part of the highway. If a road crosses a reservoir on an embankment, there are decided cases laying down that the embankment is part of the highway and it has been argued that the vesting of the embankment in the Highway Authority makes the Authority one of the owners of the reservoir with responsibilities under the Reservoirs Act 1975.[1]

There is another line of cases which suggests that, even if the retaining walls or embankments are not vested in the Highway Authority, the Authority has an easement or right of support in relation to such structures.[2] Such an easement would give the Authority the right to go on the neighbouring land and repair and maintain embankments or walls. It does not impose an obligation on the neighbouring landowner to carry out those works which are the responsibility of the Authority. If this view is right then the supporting structures are not vested in the Authority and they have no property in them other than a right of support.

## Fencing highways

The Highway Authority has vested in it the actual road and so much of the soil as is necessary to enable it to perform its statutory duties of repairing and maintaining the road as a highway. The adjoining owner retains the ownership in the subsoil and a number of other rights. Nevertheless, the Highway Authority has the right under section 80 of the 1980 Act to erect and maintain fencing along the highway for the purpose of safeguarding persons using the highway, thus separating it from the adjoining land. The Authority must exercise its power to fence so as not to obstruct any private access to premises off the highway, constructed before 1st July 1948, or

---

1  *Braintree D.C. v. Gosfield Hall Settlement Trustees* 1977.

2  *Reigate Corporation v. Surrey County Council* 1928.

built with the benefit of planning permission after that date; nor interfere with the carrying out of agricultural operations.

## The rights of the adjoining owner

There is a presumption that the owner of the land adjoining the highway owns the soil of the highway up to the middle of the road. If he owns land on both sides of the highway the presumption is that he owns the soil of the whole road. Similar principles apply to private or occupation roads. The presumption of ownership can be rebutted by evidence to the contrary. Particular care needs to be exercised in considering the ownership of recently laid out roads. If there are a number of uncompleted building plots on either side of the road, the builder has probably retained ownership over the road being built. If, however, the estate has been completed and the builder has long since disappeared it will be difficult to rebut the presumption that ownership of the road has passed to the properties on either side. If the road has been dedicated as a highway by the builder, under section 38 of the 1980 Act, ownership of the soil will still remain with the owners fronting the road. Whether the builder has dedicated the road as a highway or not, he may have been careful to retain ownership of a piece of land at the end of the road, and to exclude it from any Dedication Agreement, so that no one else in the future can obtain the benefit of using the road built at his expense in order to develop further land without buying the retained land, or a right of way over it.

In the event of a highway being stopped up and its highway status being brought to an end under statutory procedures available in the 1980 Act, or the Town and Country Planning Act 1990, then title to the soil of the road reverts to the adjoining owners unencumbered by any rights of public passage (unless the highway has been built on land which had been purchased by the Highway Authority). This reversion of ownership happens automatically as soon as a stopping up or diversion order is made and no formal procedures to transfer the land back to the adjoining owners are necessary. If the Highway Authority intend to apply for a highway to be stopped up or diverted under section 116 of the 1980 Act, they have to give 28 days' notice that they intend to apply for an order to adjoining owners and

occupiers so they may be aware of the position and
that the land will revert to them.

The owner of the soil below the highway "has the right to
and underground except only the right of passage for the
his people"[1] and to exercise all the rights of ownersh
inconsistent with the public right of passage. He can maintai   an
action for trespass against anyone laying pipes under or wires over
the highway except the Highway Authority or Statutory
Undertakers acting in pursuance of specific statutory powers. The
owner of land crossed by bridleways or footpaths can drive vehicles
along those tracks as long as he does not interfere with the public
right of passage on foot or horseback. Equally, he can maintain an
action for trespass against other people driving vehicles on public
paths across his land.

The owner of the highway can also bring actions in trespass or
nuisance to prevent encroachments on the highway, to have
obstructions removed and to stop vehicles using the highway
where there are no vehicular rights of passage. In practice the
Highway Authority has specific statutory powers to deal with
these various forms of interference with the right of public
passage and actions by adjoining land owners to deal with such
matters are rare.

## Owner's access to the highway

Although, as mentioned above, the owner of the soil beneath the
highway retains the right to dig mines and minerals, and other
miscellaneous rights of ownership are preserved, the most
important right in practice for him is the right at common law to
access to the highway at any point where his land touches it. This
right has been much restricted by the Town and Country Planning
legislation because the formation of a new means of access to a
highway, even a footpath, ranks as an engineering operation and is
development requiring planning permission. This has been
necessary with the growth in volume and speed of traffic in order
to allow a free flow of traffic along roads and to avoid dangers to

---

1   *Goodtitle d Chester v. Alker and Elmes* 1757.

D

road users arising from traffic turning on and off the road at numerous points. The Highway Authority may also make an order under section 124 of the 1980 Act to close an existing access if they consider it is likely to cause danger to, or to interfere unreasonably with, traffic on the highway. If the owner of the access objects, the order has to be submitted to the Secretary of State for Transport for confirmation. If an order is confirmed and comes into effect for closing an existing access, compensation is payable by the Highway Authority to any person suffering depreciation in the value of his property as a result. An order cannot be confirmed unless no access is reasonably required from the highway in question or unless another reasonably convenient means of access to the premises is available or will be provided.

An owner's land may adjoin the highway and his house and drive may have been built before planning permission was required under the Town and Country Planning Act 1947. Nevertheless, the matter does not end there. If there is no proper crossing from the carriageway over the footway and/or grass verge and the householder is driving cars to and from the carriageway, the Highway Authority has substantial powers under section 184 of the 1980 Act to insist on constructing a properly surfaced vehicle crossing over the footway or verge and to charge the occupier of the premises with the costs. If planning permission has been granted for the development of the land with a new access, the Authority also has power under section 184 to insist on constructing a vehicular crossing even though the premises are not yet built or occupied. Under subsection (11) of section 184 the occupier of the premises may take the initiative and request the Authority to construct a vehicular crossing but cannot insist upon them doing so. If the Authority agree to meet the request, they have to provide the occupier with a quotation for the costs of the works and then, once he has paid them the amount quoted, the Authority must make the crossing.

## The rights of public passage

Members of the public have the right to pass and repass along any

highway. They do not have the right to stop on it. As ⸜
a seventeenth century text book:[1]

> "In a highway the King has nothing but the passage foɩ
> himself and his people but the freehold and all profits
> belong to the owner of the soil."

## Direct action to deal with obstructions

On the other hand, the Courts could be equally severe about
obstructing the right of passage. In a case as early as 1630,[2] the Court
said:

> "If a new gate is erected across a highway, it constitutes
> a common nuisance even though it be not fastened and
> any of the King's subjects passing that way may cut it
> down and destroy it."

In *Reynolds v. The Urban District Council of Presteign* (1896), the
District Council, who were the Highway Authority, had grubbed up
hedges and removed post and wire railings belonging to Mr.
Reynolds because they considered that they had been erected in the
highway. They had not, first of all, tried to prosecute him for
obstructing the highway. Mr. Reynolds sued the Council for
trespass. The Court of Appeal found that the Council were entitled
to behave as they had done and dismissed the claim for trespass.
The Lord Chief Justice said:

> "I do not at all mean to dissent from some of the
> observations which were made by Counsel for the
> plaintiff, pointing out that it is desirable that the question
> of whether a private individual has encroached should
> not be decided by the public body acting in a
> high-handed way, or taking the law into their own hands.
> Where the question is one of doubt or difficulty, I think
> a judicial decision should be obtained by the public
> authority. If they proceed to act, professing to prostrate
> the encroachment under the powers of the statute,

---

1 1 Roll. Abr. 392.

2 *James v. Hayward* 1630.

without having obtained such a decision, they run a considerable risk. The burden lies on them of justifying their action; and if they fail to justify by reason of their being unable to show that there was an obstruction or encroachment, they become liable to damages for trespass. I think, however, that if they are able to show that there is an obstruction or encroachment, they have the same right as a private individual has to protect his own property. Where there is an encroachment on or obstruction to private property, the owner is entitled to remove it ...

"With respect to encroachments on or obstructions to highways, it is undoubtedly true that there are statutory provisions pointing out particular means which the local authority may adopt in order to obtain the removal of the encroachment or obstruction; but, in my opinion, those provisions are not intended to be exclusive ...

"I, therefore, come to the conclusion ... that the defendants did no more than they were justified in doing."

That right to abate a nuisance or other interference with the highway by direct action has been specifically preserved by section 333 of the 1980 Act, though obviously Highway Authorities, societies concerned with preserving rights of way and individual citizens need to be very sure of their position before they resort to such action.

## Highway to be used for passage and not for other purposes

The right to pass and repass has been narrowly construed on occasions in decided cases both in the last century and this century. In the last century it was often adjoining owners who acted to assert limitations on the rights of the public in the way they used highways crossing their land. In a criminal case in 1855,[1] it was decided that a person who went upon the highway with a gun and who attempted

---

1   *R. v. Pratt* 1855.

to shoot a pheasant which flew over him was properly convicted of trespass in search of game. One of the Judges said:

> "He said that he could not be a trespasser because it was a highway. But I take it to be clear law that if, in fact, a man be on land where the public have the right to pass and repass, not for the purpose of passing and repassing, but for other and different purposes, he is in law a trespasser."

In *Harrison v. Duke of Rutland* (1893) the Court held that the Duke's servants were entitled to hold down Mr. Harrison who was on a highway crossing the Duke's moors in Derbyshire and who was attempting to prevent the grouse flying towards the butts "by waving his pocket handkerchief, by opening and shutting his umbrella, and in other ways". In *Hickman v. Maisey* (1900) a landowner was successful in obtaining damages and an injunction against the proprietor of a racing journal who was walking backwards and forwards along a highway on the Wiltshire Downs for about an hour-and-a-half watching and taking notes of the trials of racehorses on the owner's land. These were successful actions in trespass and some lawyers have doubted whether adjoining owners can still sue in trespass since section 263 of the Highways Act 1980 has vested the highway in the Highway Authority, though this only applies to the property in the surface of the road and in so much of the soil below as is required for its control as a highway. Nevertheless, these cases have never been overruled. As will be seen later in this chapter, when the subject of allowing trading on the highway is dealt with, Parliament still pays a great deal of respect in modern legislation to the rights of adjoining owners.

Despite finding for the Duke in the 1893 case, Lord Esher, the Master of the Rolls, said in the course of his judgment that, although he had to uphold the law in favour of the Duke, he was anxious to do so in such a way as not to "interfere with the rights of the public on a highway". He said:

> "I do not think that the law is that the public must always be passing and doing nothing else on a highway. There are many things often done and usually done on a

highway by the public, and if a person does not transgress any such use and reasonable mode of using a highway, I do not think that he is a trespasser. That seems to me to show that the rights of the owner of the soil are not very large."

In this century actions on the extent of the public right of passage have usually begun at the suit of the police or the Highway Authority, often by way of a prosecution for obstructing the highway. The Courts have not found it any easier to be precise and have often themselves recognised that they are taking decisions which appear to be harsh. Leading cases have revolved around matters such as using the highway for picketing or peaceful demonstrations, rather than for watching horse-trials or interfering with grouse shooting, but the Courts have applied the same principles.

## Offences of public nuisance and obstruction

At common law, it has always been a nuisance to obstruct a highway or render it dangerous. Such nuisances are offences at common law and prosecutions can be taken by local authorities or private individuals. In addition, individuals who can show that they have suffered special damage by reason of a nuisance on the highway can recover damages from the persons responsible in a civil action.

The case of *Campbell v. Paddington Corporation* (1911) is an interesting example of the local authority themselves breaking the rules about obstruction in the highway. The Mayor, Aldermen and Councillors of the Borough resolved to erect a stand in the Edgware Road in London (at their own expense it should be said) so that they "and their friends" could watch the funeral procession of Edward VII. The stand obstructed the view from a first floor balcony of an adjoining house whose owner established that she could sell seats on her balcony for quite considerable sums to view public processions in Edgware Road. She brought an action against the Council for the loss of income amounting to £90 as nobody wanted to use her balcony once they saw the stand being erected. It was held by the Court of Appeal that the erection of the stand in the highway was a public nuisance and that the adjoining owner could successfully recover special damages from the Council of her lost income of £90.

On the other hand, a householder who claimed in 1929 that vehicles delivering goods to a rear entrance of the Mayfair Hotel in Curzon Street, London, were causing a public nuisance by obstructing his entrance next door received short shrift from the Court.[1] The Judge said:

> "The owner of premises abutting on a public highway is entitled to make a reasonable use of that highway for the purpose of obtaining access to his own premises, and of loading and unloading goods at his premises. But the right of the public is a higher right than that of the occupier, and if the use by the occupier, though reasonable so far as the particular business carried on by him is concerned, in fact causes a serious obstruction to the public, then the private rights of the occupier must yield to the public rights, and the Court will interfere by restraining the continuance of the obstruction. In every case the answer to the question whether the public right has been interfered with must necessarily depend on the extent of the use; in other words, the question is always a question of degree. This kind of question can only be determined after a careful consideration of all the facts of the case."

He went on to find that delivery vehicles to the hotel stopping for five or ten minutes in the street to unload or load goods did not amount to a public nuisance. The question, therefore, of whether the neighbour had suffered any special damage in respect of which he could bring a successful claim did not even arise.

More recently, in *Dymond v. Pearce* (1972) the parking of a large lorry for many hours on a dual carriageway was held to constitute an obstruction which resulted in a public nuisance being created because part of the highway was thereby removed from public use.

Although obstruction of the highway constitutes a nuisance in

---

1  *Vanderpant v. Mayfair Hotel Co. Ltd.* 1930.

common law, the police and Highway Authorities nowadays rely on the statutory offences created in the Highways Acts, and in particular on section 137 of the 1980 Act forbidding the obstruction of the free passage along a highway, and on prohibition of waiting orders made under the Road Traffic Regulation Act 1984, discussed in Chapter 10. Obstruction of the highway continues to be taken seriously and a police officer is empowered by statute to arrest without warrant any person in order to prevent the obstruction of a highway.[1]

## Shops extending onto the footway: street traders

The Courts have never been sympathetic, either to adjoining owners using the highway to sell articles, or to street traders in mobile vans, however wide the highway may be. It is a nuisance if an owner of land adjoining a highway uses part of the highway for the purpose of his business, for example by repairing vehicles, unless the way has been dedicated subject to the right to use it for business purposes and this is not very common nowadays. Thus in a case in 1812,[2] The Lord Chief Justice said of an adjoining owner, who was a timber merchant, that he is "not to eke out the inconvenience of his own premises by taking in the public highway into his timber yard". In a case in 1947 it was held that the sale of ice cream from a shop window causing a crowd to assemble and obstruct the highway was a wilful obstruction and that the obstruction could have been avoided by selling inside the shop in the ordinary way.[3] The sale of hot dogs from a van parked at the edge of a very wide carriageway in Oxford[4] and of fruit and vegetables from a stall in a service road which formed part of the highway in Romford[5] have been found to be obstructions and the High Court has upheld a conviction in the first case and ordered the Magistrates to record a conviction in the second.

---

1  Police and Criminal Evidence Act 1984 s.25(3)(d)(v).

2  *R. v. Jones* 1812.

3  *Fabbri v. Morris* 1947.

4  *Nagy v. Weston* 1965.

5  *Brandon v. Barnes* 1966.

Section 137 of the 1980 Act lays down:

"If a person, without lawful authority or excuse, in any way wilfully obstructs the free passage along a highway, he is guilty of an offence."

What amounts to "lawful authority or excuse" cannot be defined precisely and cases on similar facts have sometimes gone one way and sometimes the other.

The local Magistrates have sometimes wanted to be lenient to shopkeepers prosecuted for obstruction of the highway and have dismissed cases. If the prosecution has appealed by way of case stated from the local Justices, the Divisional Court have usually sent the cases back to the Justices with a direction to convict. In *Seekings v. Clarke* (1961) the Eastbourne Justices had dismissed a prosecution against a newsagent for putting articles out for sale on the pavement of a road leading to the sea. The pavement was sixteen feet wide and the defendant had taken up two-and-a-half feet of the width with his goods. The Lord Chief Justice said:

"It is perfectly clear that anything which substantially prevents the public from having free access over the whole of the highway which is not purely temporary in nature is an unlawful obstruction. There are, of course, exceptions to that. One possible exception would be on the principle of *de minimis*, which would no doubt cover the common case of a newsagent who hangs out a rack of newspapers, which, though they project over the highway, project only fractionally."

He went on to approve the comments of a Judge in an earlier case heard in Northern Ireland[1] who said:

"... it is a widespread practice for shopkeepers such as greengrocers and hardware merchants to display their goods on the footpaths in front of their shops. If the law was strictly enforced in every case, we do not see how it could be successfully contended that such practices do

---

1   *Black v. Mitchell* 1951.

not constitute an obstruction of the footpaths, rendering shopkeepers liable to proceedings under various statutory provisions ... But it is notorious that these statutory provisions are not enforced with strictness, and that proceedings are rarely brought either by the urban authorities or by private individuals ... such practices, if kept within reasonable bounds, are not likely to be regarded by the public as causing any great inconvenience, unless the street is a busy one or the footpath a narrow one. In the vast majority of cases the ordinary citizen would be apt to regard a strict enforcement of the statutory provisions as unneighbourly and unnecessary ... Nevertheless, when the matter is brought before the Court it is the duty of the Court to pronounce on the legal position."

Elected members and officers of the Highway Authority therefore have a difficult path to tread in deciding when to prosecute encroachments on the highway and when to ignore them. With greater awareness of the problems of the blind and disabled in negotiating footways, there is perhaps more understanding of the need to keep pavements clear. It will continue to be difficult to reconcile the desire of shopkeepers to increase their trading opportunities with the convenience of pedestrians who simply wish to pass along the highway. Moreover, as explained below, since 1982 it has been possible for a local Council in appropriate cases to licence trading in the highway.

As will be noticed from the judgment quoted above, the Courts have tried to draw a distinction between a temporary interference with the highway and something more permanent and have sometimes been prepared to rule that a temporary encroachment does not constitute an obstruction.

In the case involving a hot dog van in Oxford in 1965,[1] the Lord Chief Justice said:

"There must be proof that the use in question was an

---

1  *Nagy v. Weston* 1965.

unreasonable use ... it depends upon all the
circumstances, including the length of time the
obstruction continues, the place where it occurs, the
purpose for which it is done, and, of course, whether it
does in fact cause an actual obstruction as opposed to a
potential obstruction."

Despite these apparently helpful comments, from the point of view
of the hot dog seller at least, the Court went on to uphold a
conviction by the Oxford Magistrates for obstruction even though
the van was parked at the edge of a wide road, because it was
stationed there regularly each night. In an earlier case heard in
1952,[1] relating to the trunk road between London and Dover, and
the issue of whether lorry drivers could pull off the carriageway
onto the layby or verge, the Judge considered *Harrison v. Duke of
Rutland*, referred to above, and the extent of the public right of
passage. He said:

"There is no suggestion that the lorry drivers in the
present case used the road for any purpose other than for
passing and repassing, though they do stop – and why
should they not – at the café ...

"I certainly cannot find that the fact that the lorries call
at the café for refreshment causes an obstruction of the
highway. They do not obstruct the highway merely by
temporary call for legitimate and proper purpose such as
getting a meal while on the road, provided they do not
stop at a place where the mere presence of a stationary
vehicle would create an obstruction."

*Absalom v. Martin* (1973) concerned a bill poster. The local
Magistrates in Bedford dismissed a prosecution for obstruction
brought against a bill poster who drove around central Bedford,
stopped at advertisement hoardings, sometimes partially pulling his
van onto the payment, to post his bills and then moved on again.
The Divisional Court referred to *Nagy v. Weston* quoted above and
accepted the view of the Court in that case that there must be proof

---

1  *Rodgers v. M.O.T.* 1952.

that the use of the highway was unreasonable in order to convict for obstruction. They also accepted that the Bedford Magistrates, on the facts of the bill poster case, acted perfectly correctly in finding that there was no evidence of unreasonable use of the highway by the defendant having regard to his business as a bill poster and that the defendant was endeavouring to carry on his normal business in such a way as to cause the least possible inconvenience to pedestrians and other road users. They said that the Magistrates were right to acquit Mr. Martin.

Of course, in the Oxford hot dog case, Mr. Nagy was also "endeavouring to carry on his normal business" but he parked his van on the same part of the highway night after night for several hours and this was held not to be a reasonable use, though on the occasion for which he was convicted his van had only been stationary for five minutes, but he had refused to move it on when asked to do so by a Police Officer.

### Demonstrations and processions

Thus, parking on the highway for a temporary period, so long as neither an obstruction or danger is caused, is permissible. Trading on the highway is likely to amount to an obstruction contrary to section 137 of the 1980 Act. What about rights to demonstrate, make speeches, or mount a picket? In *Arrowsmith v. Jenkins* (1963) the High Court upheld a conviction for obstruction recorded against Miss Pat Arrowsmith, a prominent supporter of the Campaign for Nuclear Disarmament. Miss Arrowsmith addressed a meeting in a street in Bootle and caused a crowd to congregate to listen to her. It does not appear from the report of the case that the police asked her to stop speaking and, indeed, at the request of the police, she used her loudspeaker to ask the crowd to leave a passageway for vehicles. When she had finished speaking after twenty minutes, the police arrested her for obstructing the highway. The Court upheld the conviction recorded by the local Magistrates, saying:

> "If anybody by the exercise of free will does something which causes an obstruction, then I think that an offence is committed; there is no doubt that the appellant did that in this case."

*Arrowsmith v. Jenkins* concerned the making of a speech in the highway. *R. v. Clark* (1963) related to a demonstration in the streets of London. A number of people objected to the visit of the King and Queen of Greece in July 1963 and Mr. Clark led a crowd of between five hundred and two thousand persons through streets in the West End, starting in Whitehall and then going via Trafalgar Square and a number of other streets before he was finally arrested close to Buckingham Palace.

Instead of being charged with obstruction and dealt with summarily by the Magistrates, he was prosecuted on an indictment on a charge of inciting people to commit a public nuisance by unlawfully obstructing certain highways in London. The Judge left only two questions to the jury, first:

> "Was there a public nuisance, in fact, were the streets obstructed?"

And secondly, was Mr. Clark inciting the crowd to do this?

The jury convicted Mr. Clark and he was sentenced to eighteen months' imprisonment. The Court of Appeal said that obstruction of the streets was not a public nuisance unless it could be shown that the use of the highway was unreasonable. The Court followed an Irish case relating to a procession in Belfast in 1903, and approved the views of the Irish Judges in 1903 that many processions are perfectly lawful and that no public nuisance is created by obstruction thereby unless the use of the highway in all the circumstances was unreasonable.

The Judge had made a mistake in his directions to the jury in not leaving to them the question of whether the nuisance created was a reasonable or unreasonable use of the highway. The conviction was quashed.

It is not possible to say what the jury would have decided had they been properly directed. The fact that Mr. Clark and those following him were moving along the highways the whole time would probably have weighed in favour of Mr. Clark in the minds of the jury.

*Broome v. Director of Public Prosecutions* (1973) concerned

77

picketting during a strike of building workers and went as far as the House of Lords. A Trade Union official held a placard on the highway in front of a lorry and asked the driver to draw into the side of the road, which he did. The official asked the driver not to make a delivery to a nearby building site. His persuasion failed and the driver began to manoeuvre his lorry in order to drive onto the site. The official then deliberately stood in front of the lorry to prevent the driver proceeding. At that time an Act was in force which provided that the attendance of pickets at or near a place where a person works for the purpose of peacefully communicating information to him should not in itself constitute an offence under any enactment. Lord Reid said about this statutory provision:

> "I see no ground for implying any right to require the person whom it is sought to persuade to submit to any kind of constraint or restriction of his personal freedom. One is familiar with persons at the side of a road signalling to a driver requesting him to stop. It is then for the driver to decide whether he will stop or not. That, in my view, a picket is entitled to do."

Lord Salmon said:

> "Everyone has the right to use the highway free from the risk of being compulsorily stopped by any private citizen and compelled to listen to what he does not want to hear."

The Lords upheld the direction of the High Court that the Magistrates should convict Mr. Broome of obstructing the highway.

In 1974 there was a major political demonstration in Red Lion Square, Holborn, London, during which one person was killed and a number of persons were injured. The High Court Judge appointed to inquire into the disturbances was asked to recommend "that a positive right to demonstrate should be enacted". He said it was unnecessary because:

> "The right, of course, exists, subject only to limits required by the need for good order and the passage of traffic."

This does not seem very helpful but perhaps it serves to emphasise

the view which the Courts will take about the paramountcy
right of passage. Any activity on the highway which interferes w<sub>i</sub>.
the ability of the Queen's subjects to pass along the highway
remains likely to be punishable either under section 137 of the 1980
Act or as a public nuisance at common law.

## The powers of the Highway Authorities to maintain and protect the right of passage

Many specific rights and powers are given to Highway Authorities
by statute in order to enable them to carry out their heavy
responsibilities of maintenance. In addition to these responsibilities,
section 130 of the 1980 Act makes it explicit that "it is the duty of
the Highway Authority to assert and protect the rights of the public
for the use and enjoyment of any highway for which they are the
Highway Authority". Section 130 comes at the beginning of Part
IX of the Act which contains over fifty more sections dealing with
all kinds of obstructions, encroachments and nuisances in the
highway and giving Authorities powers to deal with acts of
interference in the highway.

There are two principal weapons given to Highway Authorities by
these fifty or so sections. Some sections create offences in respect
of which Authorities can instigate prosecutions. Other sections give
the Authorities power to serve notice on those responsible to take
the appropriate steps for the cessation of the encroachment or
interference. In the event of failure to comply, there is usually power
either for the Authorities to do the work themselves and charge those
responsible with the costs, or to prosecute for failure to comply with
the notice. Thus an Authority can require the removal of
unauthorised structures from the highway (section 143) and can
require adjoining owners to lop or cut trees and other vegetation
overhanging the highway (section 154) and remove barbed wire
fences from land adjoining the highway (section 164).

So far as offences are concerned, the most widely used section is
137 concerning obstruction of highways, discussed above. Other
sections make it an offence to place objects in the highway to the
interruption of users (section 148), to remove soil or turf from the
highway (section 131), to paint marks on the surface of the highway

(section 132) and to allow cattle or horses to stray on the highway (section 155).

## The Town Police Clauses Act 1847

Section 28 of the Town Police Clauses Act 1847 is still in force and sets out a formidable list of forbidden activities in streets. It now applies to all roads whether in the town or country. Conviction for an offence is punishable by a fine of up to one thousand pounds and prosecutions are still brought under this section. The section begins:

"Every person who in any street, to the obstruction, annoyance, or danger of the residents or passengers, commits any of the following offences ... (that is to say)"

and then sets out the list. Examples of offences are:

Every person who places any line, cord or pole across any street, or hangs or places any clothes thereon.

Every person who wilfully and wantonly disturbs any inhabitant, by pulling or ringing any doorbell or knocking at any door, or who wilfully and unlawfully extinguishes the light of any lamp.

Every person who flies any kite, or who makes or uses any slide upon ice or snow.

Every person who fixes or places any flowerpot or box, or other heavy article, in any upper window, without sufficiently guarding the same against being blown down.

Every occupier of any house or other building, or other person who orders or permits any person in his service to stand on the sill of any window in order to clean, paint...

The opening words of the section are important. If, for instance, a person flies a kite, it is necessary to show that at least one resident or passenger was obstructed, annoyed or endangered. However, in a case heard in 1938,[1] the High Court found no difficulty in

---

1  *West Riding Cleaning Co. Ltd. v. Jowett* 1938.

upholding a conviction on a window cleaning company because their employee had stood on a window sill eighteen feet above the street in order to clean the outside of the window and could have endangered passers-by underneath. He was wearing a safety belt but there was no ring outside this particular window to which he could hook the belt. The Stipendiary Magistrate who heard the case ruled that the Court was not concerned with commercial practicability and the High Court agreed with him.

## Building works – licences – scaffolding – skips

It is quite normal to see scaffolding and hoardings taking up space in the streets and to find skips for rubbish placed at the edge of the road. These intrusions in the highway are essential in order to construct and repair buildings at the edge of the highway. However they constitute obstructions and nuisances and could be treated as such. To avoid this situation Parliament has included powers in the 1980 Act for Highway Authorities to licence:

> Placing builders' skips on the highway (section 139)
>
> Erecting scaffolding (section 169)
>
> The deposit of builders' materials in the street (section 171).

The grant of a licence will protect a builder from prosecution for obstruction under section 137 because he will not be acting 'without lawful authority'. It will also defeat a claim in nuisance by owners of adjoining property who are inconvenienced by these activities in the street.[1]

Under changes introduced by the New Roads and Street Works Act 1991, Regulations can now be made by the Government by statutory instrument to enable local authorities to impose time limits on the length of time these obstructions can remain in the street and to provide for charges to be levied by local authorities if these times are exceeded.

Under section 172 of the 1980 Act, when buildings on the edge of

---

[1] *Harper v. Haden & Sons* 1932.

the street are being erected, repaired or demolished, the builder must erect hoardings to the satisfaction of the Highway Authority. The Authority can also require the builder to make a convenient covered platform and handrail to serve as a footway for pedestrians outside the hoarding and can require the hoardings to be lit at night time.

## Trading in the highway

The Authority's property in the highway and its powers do not extend further than is necessary for the maintenance and use of the highway. For instance, prior to the enactment of section 87 in the Public Health Act 1936, they could not erect public conveniences in the highway. Except under specific statutory authority they have no power themselves to obstruct the highway or erect obstructions in it. Until a few years ago they could not give permission to anyone else either to erect any form of structure in the highway or to trade in the highway.

In 1982 a Local Government Miscellaneous Provisions Act conferred powers on Authorities to carry out activities or permit other people to carry out activities on the highway which are not strictly necessary to enable the public to use the highway for travelling. The 1982 Act wrote into the 1980 Highways Act sections 115A to 115K. Both the Highways Authorities and the District Councils now have powers to provide and operate facilities for recreation and refreshment within the boundaries of the highway. They can also give other people permission to locate and operate such facilities within the highway. This can include allowing restaurants to put tables and chairs on the pavement and permitting the erection of shopping booths.

In future therefore Councils can lawfully permit hot dogs to be sold from mobile vans and can charge fees for these concessions. Significantly however they cannot themselves provide refreshment facilities which will result in the production of income or allow other people to do so unless they have first obtained the consent of the frontagers. This is a recognition of the rights of the frontagers and the owners of the sub-soil. Also section 115F(2) provides that "except where the Council are the owners of the sub-soil" the charges they can make for selling refreshment or trading concessions shall not exceed the cost of providing the facilities and such charges as will reimburse the Councils their reasonable expenses in making the arrangements. Councils also have powers under these provisions to provide centres for advice and information within the highway and to permit the erection of advertisement hoardings.

Sections 115A to 115K apply to highways from which vehicular traffic is excluded by a Traffic Regulation Order, or by their nature as footways or walkways. Thus, permission cannot be granted to place chairs and tables in the carriageway of a highway.

# Chapter 6

# Maintenance of Highways and Bridges

There are no Statutes or Regulations prescribing the standards to which Highway Authorities should maintain highways.[1] Indeed, highways vary so much in their size and importance and in the volume of traffic which they carry that it would be difficult to lay down precise standards. Section 58 of the 1980 Act requires a Court to have regard to the character of the highway, the traffic which could reasonably be expected to use it and the standard of maintenance "appropriate for a highway of that character and used by such traffic" in determining whether or not a Highway Authority has carried out the duties to maintain the highway with sufficient care as to enable it to escape liability for an accident which may have occurred on that road. In one of the leading cases on the meaning of the duty to maintain, referred to in Chapter 4, Lord Justice Sachs said the duty "is reasonably to maintain and repair the highway so that it is free of danger to all users who use that highway in the way normally to be expected of them".

The Authority must consider not only the danger of personal injury to persons using their highways but also the possibility of damage to the vehicles using the road. Thus, in the Wiltshire case, referred to in Chapter 3, the Court found that the road was in such a state as to be impassable to Milk Marketing Board lorries coming to collect milk from the farm on the road and, therefore, there had been a failure in the duty to maintain. The Judge said:

> "The tanker was withdrawn because if it continued to use the road it would be damaged ... The road did not

---

1  The Department of Transport does issue papers recommending certain standards in the design of various types of roads.

have to be totally impassable. If its condition gave rise
to the risk of excessive damage to a vehicle that was
entitled to use it ... the road was dangerous ... "

## Distinction between maintenance and repair

Maintain has a wider meaning than repair and the Highway
Authority has the duty to deal with snow and ice on the road
rendering its surface dangerous to users even though it is winter
weather rather than a failure to discharge repairing obligations
which has caused these conditions. In *Hereford and Worcester
County Council v. Newman*, referred to in Chapter 3, a distinction
was also drawn between (a) the surface of a highway being out of
repair, (b) encroachments onto the highway by overhanging
vegetation and (c) obstructions caused by the erection of fences
across the highway. In Newman's case, the Court held that allowing
vegetation to grow in the surface of the highway was a failure to
keep the road in repair but that failing to deal with fences erected
in the road and overhanging vegetation did not amount to a failure
to keep the road in repair. In practice, this does not make a great
deal of difference to the Highway Authority because they have a
duty to assert and protect the rights of the public to the use and
enjoyment of the highway under section 130 of the 1980 Act. As
was discussed in the last chapter, they have ample powers to deal
with any form of encroachment, obstruction or interference with
the highway and they would probably regard their work in dealing
with these interferences as coming under the general umbrella of
their maintenance responsibilities.

## Drainage

The vesting of the surface of a highway in the Highway Authority
under section 263 of the 1980 Act was discussed at the beginning
of Chapter 5. Under section 264 the drains "belonging to a road"
also vest in the Authority together with the right to continue to use
any other drain or sewer which has been used in the past in
connection with the drainage of the road.

The cases of *Burnside v. Emerson* and *Tarrant v. Rowlands*, referred
to in Chapter 4, illustrate that the Authority's duty to maintain

includes a duty to prevent water gathering on the surface of the highway, thus endangering drivers and their vehicles. Section 100 of the 1980 Act gives the Authority powers to drain roads and prevent surface water flowing onto them. They can construct in the highway, "or in land adjoining or lying near to the highway" such drains as they consider necessary and they can scour, cleanse and keep open all drains situated in the highway or in adjoining land. Drainage is accepted as one part of the normal maintenance responsibilities of the Authority. The Authority can also fill in a roadside ditch if the adjoining occupier agrees, or pipe the ditch even if the occupier does not agree and thereafter fill it in. Compensation is payable to an owner or occupier for any damage done in the exercise of these powers (section 101, 1980 Act).

At common law the owner of higher land is not liable for water running off his land on to lower land if the water has gathered on his land naturally, for example rain water. However, if he changes the configuration of his land or surfaces it, he will be liable for damage caused by the run off of water flooding on to his neighbour's land. These principles apply to Highway Authorities and therefore they must make arrangements to drain their highways and not allow the water to flood on to adjoining land.[1]

### Grass cutting

Similarly, the responsibility of the Authority to keep roads safe as part of its maintenance duties requires Authorities and their contractors to engage in an extensive programme of grass cutting on highway verges during the spring and summer months in order to maintain visibility for drivers, particularly for those emerging at junctions on to busy roads.

### Trees – liability of adjoining owners

In a number of cases, landowners have been held liable in nuisance when trees rooted on their land have fallen across a highway and caused injury to road users. A landowner has a duty to take care to prevent danger to users of the highway from trees on his land. If he

---

1 *Thomas v. Gower Rural District Council* 1922.

wishes to avoid liability for accidents, he should institute a regular system of inspection by experts, assuming that he does not have the necessary knowledge himself, of the trees within falling distance of the highway so that he is advised of any tree that has become diseased or unstable.[1] A landowner may not be liable for nuisance in respect of a branch of an apparently healthy beech tree growing on his land which suddenly breaks off due to a latent defect, which could not have been discovered by inspection and falls upon a car passing along the road.[2]

In *Quinn v. Scott* (1965) the High Court considered an accident on the A614 road passing through National Trust Land at Clumber Park, Derbyshire. There was a belt of beech trees 200 years old on the National Trust estate running for about a mile along the road. One of the trees, 90 feet tall, and 33 feet from the carriageway, fell across the road, resulting in a collision between two vehicles. The Judge found that there was an appearance of unhealthiness in the tree indicated by the thinness of the foliage and signs of die-back. It was old and there was no protection in its position from winds blowing from the West. It was close to a fast main road. In these circumstances the Judge said:

> "I am clearly of the opinion that the reasonable landowner would say, or ought to say: 'I will not take the risk. There is no assurance that other signs of decay will appear before the tree falls, and if it falls across that road the consequences may well be disastrous. I will not wait. Let the tree be cut down at once'."

Taking that view he found the National Trust liable to pay damages to the motorist.

### Powers of Highway Authority to deal with dangerous or overhanging trees

The Highway Authority has powers under section 154 of the 1980 Act where a tree overhangs a highway or is dead or diseased so as

---

1  *Caminer v. Northern and London Investment Trust* 1950.

2  *Noble v. Harrison* 1926.

to endanger or obstruct the passage of vehicles or pedestrians to serve notice on the owner of the tree, or on the occupier of the land on which it is growing, to require him to lop or cut it within 14 days of the service of the notice. A person aggrieved by such a requirement can appeal to local Magistrates. If there is no appeal, or if the Magistrates dismiss the appeal, and if the owner or occupier fails to comply with the notice, the Highway Authority can carry out the necessary work and recover the expenses from the owner or occupier.

Section 154 does apply to trees standing well within private land which are tall enough to fall across the highway. It does not contain any emergency power for the Authority to enter on the land and cut down the trees immediately. Fourteen days' notice has to be served on the owner requiring the trees to be felled, and if there is an appeal against the notice, entry on the land to do the work cannot be made until the appeal is disposed of. If the Highway Authority are worried that the trees will fall in the meantime, all they can do is to use their emergency Traffic Regulation powers discussed in Chapter 10 to close the road and divert the traffic onto other roads.

The power contained in section 154 is discretionary and it would be an impossible task for Highway Authorities to inspect every tree along every highway in their area. It could be that they do have definite knowledge of the dangerous state of a tree along one of their highways and choose not to exercise their powers under section 154. There appear to be no reported cases of accidents where a highway user has tried to sue the Authority, as well as the owner of the tree falling on the road, on the basis that the Authority should have exercised their powers under section 154 to have the tree dealt with.

### Trees growing in the highway

As explained in the beginning of the previous chapter, the subsoil of roads, which have been dedicated as highways, remains with the adjoining owners. Dedication may have taken place when trees were already growing in the highway, for example along the highway verges. The roots of the trees are in the subsoil and arguments have taken place as to whether the adjoining owner or the Highway Authority is responsible for damage caused by the trees. In one case[1] there was a dispute as to whether the tree was actually in the highway rather than in the hedge and counsel for the defendant occupier argued that if the tree was growing in the highway, the Highway Authority controlled the tree and could remove it. The Judge did not accept that, even if the tree was in the highway, the adjoining owner could escape liability. He pointed out that the subsoil remained in the ownership of the adjoining owner as did "the growing produce, whether grass or trees, on such part of the highway as is not being used for traffic; in other words, the verges". He continued:

> "For my part, the fact that the Highway Authority, and the Minister (the road was a trunk road), can do this, or may have the duty to do it, namely to remove obstructions, does not necessarily exonerate the adjoining owner, in whom the property in the tree still remains. It seems to me that he may be liable in certain circumstances for nuisance."

These comments should perhaps be treated with some caution because, in *Stillwell v. New Windsor Corporation* 1932, another Judge had taken the opposite view and decided that trees in the highway vested in and were under the control of the Highway Authority. The Judge considered the vexed question of exactly what vests in the Highway Authority and said:

> "In my view, for all the purpose of exercising the rights of the Highway Authority, these trees are to be treated as the Highway Authority's trees, and, if they think it convenient to remove them, it is proper that they should remove them. I am not called on in this action to decide to whom the timber would belong when the trees were removed."

Since in a much more recent case in 1984[2] the Highway Authority was held liable for damage done to a nearby house by the roots of

---

1  *British Road Services Ltd. v. Slater* 1964.

2  *Russell v. Barnet L.B.C.* 1984.

a tree growing in the highway, the judgment in Stilwell's Case would be logical, as otherwise the Highway Authorities could not take action to deal with trees even though they can be held liable for the damage caused by those trees.

## Trees: power of Highway Authority to plant

Under sections 64 and 96 of the 1980 Act, a Highway Authority can plant trees within the boundaries of a highway, either for ornament or in the interests of safety, for example to demarcate land lying between the carriageways of a dual carriageway. If a tree planted by the Highway Authority within the highway falls and causes an accident, the Authority will be liable for damages. In *Hale v. Hants and Dorset Motor Services Limited* (1947) the plaintiff was riding as a passenger on the top of a bus when the branches of a tree struck the windows of the bus and a piece of glass hit the plaintiff in the eye, which had to be removed. The tree had been planted on the side of the highway by the Poole Council. Counsel for the Council tried to argue that the Highways Acts specifically authorised the Council to plant trees in the highway and that they were not responsible if the trees grew and branches began to overhang the carriageway. The Court rejected this argument and awarded damages against the Council.

Subsection (3) of section 96 of the 1980 Act, in conjunction with sections 246 and 282, provides a useful power in connection with new highways and highway improvement schemes for the Highway Authority to acquire compulsorily slightly more land than they need for a new highway so that they can plant trees and lay out grass on it with a view to screening the highway and making it more attractive.

## Street lighting

Highway Authorities have the power to provide lighting for the purposes of any highway under section 97 of the 1980 Act. They may delegate their lighting functions with respect to any particular highway or highways to District Councils or Parish Councils and pay them to carry out the work on an agency basis. These Authorities are also lighting authorities under other legislation in their own right and retain powers to provide street lighting systems out of their own funds, though these powers can only be exercised with the consent of the Highway Authority. In practice, a Highway Authority will light roads to prevent danger to road users but will not be concerned with the need for street lighting for other reasons such as the prevention of crime. In a case heard in 1921[1] it was held that the Authority has no duty to light the highway, even dangerous places on the highway, unless it has itself created the danger, for example by excavating the street and leaving the hole without light. This case has never been overruled. In view of the abolition of the rule in 1961 that the Highway Authority is not liable for non feasance or omissions to perform its duty to maintain the highway, a similar case might be decided differently today, particularly if it did concern a well known danger spot for road users.

## Street furniture

Highway Authorities have a wide range of miscellaneous powers to provide street furniture of various types or to allow other bodies to provide such equipment. Under the Public Health Acts, the District Councils can erect and maintain seats and drinking fountains for the use of the public and troughs for watering horses or cattle. Again, District Councils and public service vehicle operators can provide bus shelters within the highway so long as they have the consent of the Highway Authority. The District Council can also erect statues or monuments in the highway if they have the consent of the Highway Authority. The Highway Authority, and the District Council with the consent of the Highway Authority, may provide and maintain litter baskets and other receptacles for refuse and bins for the storage of sand and grit. It is interesting to note that section 185 of the 1980 Act which grants this power provides in subsection (4):

> "Nothing in this section is to be taken as empowering an Authority to hinder the reasonable use of a street by the public or any person entitled to use it or as empowering

---

1 *Sheppard v. Glossop Corporation* 1921.

an Authority to create a nuisance to the owner or occupier of premises adjacent to a street."

Yet again, the paramountcy of the public right of passage is underlined and emphasis is also placed on not causing a nuisance to adjoining occupiers or frontagers.

## Road signs

As part of the Highway Authority's duty to maintain the highway under section 41 of the 1980 Act, they must see that the highway is free from danger to users of the highway. One of the criteria in deciding whether an Authority can take advantage of the special defence in section 58 of the 1980 Act, discussed in Chapter 4, is that if a highway is out of repair, warning notices of its condition have to be displayed. The Authority also has extensive powers under section 65 of the Road Traffic Regulation Act 1984 to erect traffic signs warning drivers of hazards ahead on the roads, such as junctions, sharp bends, slippery surfaces, uneven roads, steep hills and roadworks. If an Authority does not make proper use of these powers, it is failing in its duty to maintain the highway. In *Bird v. Pearce* (1978), Mrs. Bird was a passenger in a car driven by her husband. Their car collided with Mr. Pearce's car, which was emerging from a minor road. Both drivers were driving too fast and ten per cent of the damages were apportioned against Mr. Bird, who was driving along the main road. Ninety per cent was apportioned against Mr. Pearce, who then joined the Somerset County Council in the action on the basis that they had not given him adequate warning that he was approaching a junction with a major road. Normally there were double white lines across the minor road at its junction with the major road, preceded back up the minor road by long white dashes. However, a month before the accident occurred, the minor road had been resurfaced and the markings had been obliterated. The Judge found that the County Council were partly responsible for the accident because they had not erected a temporary sign indicating the proximity of the junction and he ordered the Council to contribute one-third of the ninety per cent of the damages awarded to Mrs. Bird, which Mr. Pearce had to find.

92

## Maintenance of grass verges

Given the enormous mileage of highways to be maintained by Authorities and the fact that they do not have infinite resources, it is helpful to Authorities if adjoining owners can be encouraged to look after the grass and plant low shrubs on the verges outside their properties between their boundary and the metalled footway or carriageway. Owners will themselves often ask to be allowed to do this. Until relatively recently it was not possible for Highway Authorities to permit this and indeed they were under a duty to take action to deal with what could be regarded as obstructions or unlawful encroachments. The 1980 Act contains power in section 142 for the Highway Authority to grant licences to adjoining owners to maintain and plant trees, shrubs, plants or grass in the highway. No more than a terminable licence may be granted. The licence can be personal to the licensee only or the Authority can annex it to his adjoining property. In that case when the property changes hands the licensee must notify the Highway Authority of the change but it is not necessary to issue a new licence.

## Bridges, tunnels, embankments, retaining walls

Where a highway passes over a bridge, the bridge is part of the highway and the normal duty to maintain under section 41 of the 1980 Act applies. If the highway is in a cutting, the Authority will have to maintain the sides of the cutting and prevent soil falling onto the surface of the road. Where a highway is supported or flanked by retaining walls, those walls are likely to be maintainable by the Authority as part of its duty to maintain the highway whether or not the abutments and walls are vested in the Authority. As explained in the previous chapter, even if they are not vested in the Authority, the Authority will have an easement or right of support in relation to the structures and can enter onto neighbouring land to carry out its maintenance responsibilities. It was held in *Sandgate U.D.C. v. Kent County Council* (1898) that, if a highway runs along the seashore, an embankment, seawall and groynes may be necessary for its protection and in that case will be repairable with it by the Highway Authority. In the case of *Reigate Corporation v. Surrey County Council* (1928) the High Court considered responsibilities with regard to tunnels through which highways pass. The case concerned a road running through Castle Hill, Reigate. The Judge said:

> "I am prepared to hold, and do hold, that the walls and roof of the tunnel which were erected simultaneously with the making of the road, and which operate to keep, and are necessary to keep, the surface of the road free from that which would or might otherwise obstruct it, may properly be said to form part of the roadway.

> "If I am wrong in this, another view is open. If the walls and roof are not part of the road, yet they may constitute something, apart from the road, which it is necessary to maintain and repair for the purpose of maintaining and repairing the road; something which, if allowed to get into disrepair, may bring about the obstruction and disrepair of the highway. In other words, that the case is analagous to the Sandgate case, the repair and maintenance of the walls and roof of the tunnel correspond to the repair and maintenance of the seawall and groynes in that case."

Under section 92 of the 1980 Act a Highway Authority may reconstruct a bridge which is a publicly maintainable highway either on the same site or on a new site within 200 yards of the old one.

## Railway and canal bridges

Special rules govern bridges carrying highways across railways and canals. These rules are set out in sections 116 to 122 of the Transport Act 1968. Until 1968, many of these bridges belonged to British Rail, London Transport Underground or British Waterways. They had been constructed in the nineteenth century when the railway or the canal had cut through the road and the responsibility for maintenance rested on the railway or canal companies. Under section 116 of the 1968 Act, responsibility for the surface of the highway over the bridge and over the approaches to the bridge was transferred to the Highway Authority.

Under subsection (6) of section 116, the Boards of British Rail,

London Transport and British Waterways remained responsible for the structure of the bridges and the approaches. Consequently, the Act provides that the Highway Authorities are not to be responsible for any defect in the surface of the highway so far as it is attributable to the failure of any of the Boards to discharge their responsibility with regard to the maintenance of the structure. For their part, the Highway Authorities must not, without the consent of the Boards concerned, increase to a significant extent the weight of the materials constituting the surface of the highway.

In constructing a new bridge, the Boards must secure that it has the required load bearing capacity for existing traffic. Difficult problems arise however with regard to old bridges, which are not strong enough to carry lorries, because lorries are now much heavier than when the bridges were first built. From 1st January 1999 forty tonne lorries will be permitted on the English roads, as already permitted in other parts of the European Community, and a major programme of strengthening highway bridges and structures is now in progress. Although the point is not quite settled, the necessary monies for this will probably have to be found by the Highway Authorities and not the Boards.

### Privately maintainable bridges

There are still some privately maintainable bridges. These have come into existence where an individual has for his own private purposes created the necessity for a bridge, for example by cutting a private road, canal or dyke across the highway and has built a bridge in order to enable the public to continue to exercise their right of passage. Use by the public, being essential if they wanted to continue to use the highway, cannot make the bridge maintainable at the public expense.[1] There are provisions in sections 93 to 95 of the 1980 Act for the Secretary of State in the case of trunk roads, and the local Highway Authority in the case of other roads, to enter into agreements with the owners of privately maintainable bridges for the transfer of the property in the bridge and responsibility for its improvement and maintenance.

---

1   *R. v. Isle of Ely Inhabitants* 1850.

These agreements will normally contain financial provisions. If the bridge has become an expensive liability to the owner, he may be willing to pay the Highway Authority a capital sum in order to rid himself of the responsibility of maintenance. If, on the other hand, it is a bridge for the use of which the owner has the right to charge tolls by virtue of a Royal Charter or Act of Parliament, then the Highway Authority may have to buy out the owner of the bridge and there is provision for this under section 271 of the 1980 Act. In that event, the Authority can continue to charge tolls, if they wish, or can cease to exercise this right.

In the event of failure to agree on future responsibilities under section 94 of the 1980 Act, either the owners or the Highway Authority can apply to the Secretary of State for an order under section 93 of the Act. Such an order can require owners or the Highway Authority to reconstruct the bridge and can require ownership of the bridge to be transferred to the Highway Authority.

### Bridges over highways and underpasses

The Highway Authority is not responsible for maintaining bridges over the highway, or underpasses under it, unless the bridges or underpasses are themselves highways carrying traffic, or rights of way, or have been provided by the Authority for pedestrian safety. In other cases, where an owner of land on either side of the highway wants to build a bridge over it, or an underpass under it, he can do so but he must obtain a licence from the Highway Authority to construct the bridge or underpass. In the case of a bridge, the procedure is laid down by section 176 of the 1980 Act. It shall be a condition of every licence that the person to whom it is granted shall remove or alter the bridge if at any time the Highway Authority consider the removal or alteration necessary or desirable in connection with the carrying out of improvements to the highway. A licence granted under section 176 cannot authorise any interference with the convenience of persons using the highway and no fee can be charged for granting a licence except a reasonable sum to cover administrative expenses.

The position with regard to underpasses is similar. The power to permit tunnelling or boring under the highway is now contained in section 50 of the New Roads and Street Works Act 1991, discussed in Chapter 9.

E

Chapter 7

# Improvements to Highways and Construction of New Roads

The enormous increase in car ownership since the Second World War, the increase in the size and weight of heavy goods vehicles and the growth in their numbers have all made heavy demands on roads. Even recently constructed motorways are having to be repaired and reconstructed within a few years of opening. The pressure on Highway Authorities to increase the strength and capacity of roads or to build new roads and bypasses has been continuous for many years now. It has not been necessary to pass new legislation to give Highway Authorities further powers because the Highways Acts have always contained wide powers both to improve existing roads and to build new roads. Section 24 of the 1980 Act enables both the Secretary of State and the local Highway Authority to construct new roads. The Government has also investigated the possibility of attracting private finance into road building and the New Roads and Street Works Act 1991 enables new roads to be entirely financed by private companies who can be authorised to charge tolls for their use in order to recoup their expenditure – not unlike the eighteenth century turnpike roads.

## Improvements to highways

Part V of the 1980 Act, sections 62 to 105, confers a substantial number of powers on Highway Authorities to carry out improvements. Section 62 states that the Authority can carry out "any work (including the provision of equipment) for the improvement of the highway". This is followed by a number of sections conferring specific powers. For example, section 64 provides that, where the highway consists of a made up carriageway, the Authority may construct a dual carriageway and works at crossroads for regulating the movement of traffic, such as roundabouts. It goes on to say that the powers conferred to carry out these works include power to light the works, pave or grass them, erect rails or fences on or around them and plant trees, shrubs and other vegetation, either for ornament or in the interests of safety. Section 65 enables the Authority to construct a cycle track by the side of any highway which comprises a made up carriageway.

## Safety of pedestrians

There follow a number of powers dealing with the safety of pedestrians. Section 66 lays down that it is the duty of a Highway Authority to provide a footway by the side of a highway which comprises a made up carriageway "as part of the highway in any case where they consider the provision of a footway as necessary or desirable for the safety or accommodation of pedestrians". The Authority can also provide and maintain safety barriers and rails. Section 68 enables the Authority to place island refuges on the highway to assist pedestrians crossing the carriageway. Sections 69 and 70 authorise Authorities to provide subways and footbridges. It is interesting to note that with regard to both subways and footbridges, the Highway Authority can maintain them but that it is also permitted to "alter, remove or close temporarily" such subways or bridges. Although they are part of the highway members of the public do not have the right to use them for all time and the strict procedures described in the final chapter of this book for stopping up highways are not applicable.

## Surfacing highways

Under section 99 of the Act, a Highway Authority has the power to convert any highway maintainable at the public expense into a metalled highway and under section 104 they can, in relation to such a highway, treat it for mitigating the nuisance of dust. Although it hardly seems necessary to give power to metal a highway, this provision does overcome the difficulty of landowners arguing that they have dedicated a road as a highway in its existing state and the public must take it as they find it.

## Widening highways: altering levels

A Highway Authority may widen any highway for which they are the Authority under section 72 and may for that purpose agree with the person having power in that behalf for the dedication of the adjoining land as part of the highway. That will not always be easy and later in this chapter local authorities' powers to acquire land compulsorily to improve highways will be discussed. The Authority can level a highway under section 76 and they can also raise or lower the level of a highway under section 77. If, however, they do alter the levels, they may have to pay compensation to any person who sustains damage by reason of the execution of the works – for example because the entrance to his drive is then several feet above or below the level of the road.

## Improving visibility at junctions

The Authority can execute works for cutting off the corners of a highway, assuming that they have sufficient land within the highway. There is also a useful power in section 79 of the 1980 Act to require adjoining occupiers to lower walls or vegetation to a maximum height stated by the Authority and not to erect new walls above that height or to allow vegetation to grow above that height. The power can be exercised if the Authority deem it necessary for the prevention of danger arising from obstruction to the view of persons using the highway with respect to any land at or near any corner or bend in the highway or any junction of the highway with a road to which the public has access. The power is useful because, in effect, it allows owners to keep their land and the Authority to secure a visibility splay without having to acquire land to throw into the highway. Owners or occupiers who can show that they have sustained a loss in consequence of requirements to lower their fences or walls, or who can prove that their property has been injuriously affected, are entitled to compensation for the injury sustained. If all that is required is that the owner should reduce his front fence by two or three feet, the compensation which can be claimed is not likely to be great.

## Cattle grids

Highway Authorities can provide cattle grids in the highway where it is expedient to do so for controlling the passage of animals along the highway. If they do provide a cattle grid, they also have to provide, by means of a gate or other works, facilities for the passage under proper control of animals and all other traffic that is unable to pass over the cattle grid. Where a highway has been dedicated subject to a right to place a gate across it, an Authority, having provided a cattle grid, can require the gate to be removed under section 86 of the Act. The Authority may also enter into agreements with persons who wish cattle grids to be provided for those persons to make a contribution towards the cost of providing and laying the cattle grid.

## Highways over or under rivers and canals

Orders made by the Secretary of State, or by a local Highway Authority and confirmed by the Secretary of State, can make provision for the construction of a bridge over or a tunnel under any specified navigable waters. Orders can also be made authorising a Highway Authority to divert part of a navigable watercourse if it is necessary or desirable in connection with the construction or improvement of a highway. Non-navigable watercourses can be diverted by Highway Authorities in connection with the construction or improvement of a highway without the need for a formal order. However, twenty-eight days' notice of intention to divert the watercourse must be given to owners and occupiers who will be affected and, if they object, the local Highway Authority cannot carry out the works before the Secretary of State has given his consent.

## Picnic sites

Under section 112 of the 1980 Act the Secretary of State for Transport can provide picnic sites, public conveniences and restaurant facilities on land adjoining trunk roads. The powers of a local Highway Authority are not so wide but they can provide public conveniences on or adjoining the highway. Both the County and the District Councils have powers elsewhere in the Countryside Act

1968, section 10, to provide picnic sites and car parks for motorists alongside highways.

## Road humps and traffic calming

The Transport Act 1981 inserted a new power in sections 90A to 90F in the Highways Act for an Authority to construct road humps in a highway maintainable at the public expense if the highway is subject to a speed limit of thirty miles per hour or less or if the humps are specifically authorised by the Secretary of State. Where a road hump is constructed under this power, section 90E provides that it is not to be treated as constituting an obstruction to the highway but as "part of the highway". The result is that the Authority has to maintain the hump as part of the highway and statutory undertakers exercising powers to break open the highway have to reinstate the humps. The Authority can remove the humps at any time it no longer wishes to maintain them.

These new provisions have been warmly welcomed by householders living in quiet residential roads who wish to slow down passing traffic, but less warmly welcomed by those same householders when they are passing along other people's roads. It is slightly odd that Parliament has inserted this section in a part of the 1980 Act dealing with 'Improvement of Highways'. Humps can, however, only be laid down in accordance with Regulations made by the Secretary of State under section 90D and he is empowered to make such provision in the Regulations "as appears to him to be necessary or expedient in the interests of safety and the free movement of traffic". He can also provide that road humps should only be constructed in certain types of highways and can impose requirements with regard to the dimensions and spacing of road humps. Regulations made so far have been tight[1] and local Highway Authorities have to explain to many residents that the Regulations prevent the laying of humps in their particular road.

The Traffic Calming Act 1992 has added further subsections to section 90 enabling Highway Authorities to construct traffic calming works. These works are to be of descriptions prescribed by

---

1  S.I. 1990 Nos. 703 and 1500.

the Secretary of State in Regulations yet to be made. Some Authorities do not like humps because of the risk of accidents and this Act can be used to develop other methods of slowing down traffic.

## Contributions by developers towards highway works

Section 278 of the 1980 Act is a power widely used by Highway Authorities to obtain financial contributions towards the cost of new highway works. It enables the Authority to enter into an agreement with any person who would derive a special benefit if highway works incorporated particular modifications, additions or features or were executed at a particular time. In the case of a major new development, such as a new out of town supermarket or a new industrial estate, the Highway Authority will anticipate that there will be an enormous increase in traffic on the road fronted by the new development and will wish to ensure that a safe means of access to and from the existing highway is provided in the development and that problems are not caused or exacerbated in the existing road network. They may ask for substantial works within the existing highway, such as the construction of a new roundabout or a right turning lane or deceleration and acceleration lanes, or may require a new length of highway to be built. The Highway Authority may object to planning permission being granted for the new development unless it can be certain that these works will be carried out. In these circumstances, a developer will often be willing to pay an Authority the cost of the Authority executing the necessary works so that it can obtain planning permission for its development. In the case of motorways and trunk roads for which the Secretary of State for Transport is the Highway Authority, he can actually direct refusal of a planning permission or insist on conditions being imposed on the permission requiring highway works to be carried out before the new development is opened.

If any person enters into a Section 278 Agreement and fails to pay the financial contribution agreed, the Highway Authority may direct that any new means of access or other facility afforded by the works shall not be used until the money is paid. They can also recover the amount due from the owner of any land benefitted by the works and declare the amount due to be a charge on the land registerable as a local land charge.

Any works executed under a Section 278 Agreement must be works which the Highway Authority could themselves have carried out under their powers to construct new highways or improve existing ones.

### Construction of new highways – statutory procedures

Section 24 of the 1980 Act contains a general power for the Secretary of State and local Highway Authorites to construct new highways. The building of new roads is not universally popular and heated public controversy as to whether a particular new road is necessary, and as to the route over which it should run, is common. There are important statutory procedures to be observed in building new highways and a period of years will be necessary between initial planning of any new road and the contractor arriving on site to start moving earth and building the road. In the case of a new trunk road, the Secretary of State will publish a draft order under section 10 of the 1980 Act setting out the route for the new road and, in the case of a new motorway, he will publish a draft scheme under section 16. If there are any objections a local inquiry will be held and the Secretary of State must consider the report of the Inspector before deciding whether or not to make the order or scheme. In the case of other roads the local Highway Authority will submit an application for planning permission for the construction of a new road along a route shown on the application. Although the local authority can give itself planning permission it is usually the case with any major road scheme that there are objections and the Secretary of State for the Environment will decide to call in the application for his own determination. He too will cause a local inquiry to be held and will consider the report of his Inspector before deciding whether or not to grant planning permission to the local Highway Authority.

The construction of the new road often cuts across existing roads and, if the new road is intended to be a major through route, the authority promoting the scheme, be it the Secretary of State or the local Highway Authority, will usually also need to obtain authority to stop up or divert the existing minor roads. Often this will be done by altering their levels to take them under or over the route of the new road. Orders under section 14 or 18 of the 1980 Act will be made by the local Highway Authority and submitted to the Secretary of State for confirmation or will be prepared in draft by the Secretary of State, and then made by him, in both cases after a local inquiry if there are any objections. Under section 125 of the 1980 Act these side road orders can also include provision to stop up private means of access to premises adjoining the roads, provided no access is reasonably required or another reasonably convenient means of access to the premises concerned is made available.

## Acquisition of land – compulsory purchase

Under Part XII of the 1980 Act, Highway Authorities are given wide powers to acquire land by agreement or compulsorily.

There is often insufficient land within the highway to enable widening of the carriageway, the construction of a new footway or other improvements to take place. In order to improve a road or construct a new road, it is necessary to acquire relatively small areas of land from a large number of different owners along the line of the proposed road. Even if some owners are willing to sell land by agreement to a Highway Authority in order to allow widening or the construction of a new road, there will usually be as many, or more, who will not sell unless compelled to do so. Extensive powers of compulsory purchase have, therefore, been given to the Secretary of State for Transport and local Highway Authorities in Part XII of the 1980 Act. Sections 238 and 239 enable the Secretary of State or the local Highway Authorities to acquire land compulsorily for the construction or improvement of a highway. Section 240 enables the Secretary of State to acquire land to provide a trunk road picnic area and the local Highway Authority to acquire land to provide public conveniences on or adjoining the highway. It is also wide enough to enable the Highway Authority to include land close to the existing or proposed highway in the Compulsory Purchase Order for working space because subsection (2) refers to the acquisition of land required "in connection with the construction or improvement of a highway".

The Secretary of State publishes draft Compulsory Purchase Orders for roads where he is the Highway Authority. Orders of local Highway Authorities have to be submitted to him for confirmation. In both cases, if there are objections from owners of property affected, there will be a local inquiry before the order is made or confirmed by the Secretary of State. The procedures for compulsory acquisition are lengthy and Highway Authorities need to bear this in mind when planning their programmes of road construction and improvement.[1] It will usually take anything between one and three years from the time a Compulsory Purchase Order is made to the time it is confirmed.

Where the Highway Authority does not need to own land but simply to have a right to do something over or under it, such as the right to construct a bridge carrying a highway over land, they can make an order for the compulsory creation of a right to be granted over the land in the nature of an easement. The owner of the land will still be able to claim compensation for the depreciation in the value of his land, though the compensation payable should be less than if the freehold had been acquired.

Under section 246, the Authority can acquire compulsorily land needed for the purpose of mitigating any adverse effect which the existence or use of the highway to be constructed will have on the surroundings of the highway. For example, land can be acquired and used at the side of the highway to plant shrubs and trees to muffle the noise of traffic and to conceal the road. This power cannot be exercised compulsorily after the highway has been opened to traffic. Except in certain special cases, the Highway Authority cannot include in a Compulsory Purchase Order any land more than two hundred and twenty yards from the centre line of the proposed new road.

## Compensation

A landowner whose land is taken by compulsory purchase will receive by way of compensation the value which his land could have been expected to reach if sold on the open market. He will also be

---

1   The principal statute on the procedures to be followed is the Acquisition of Land Act 1981.

entitled to compensation for injurious affection to the remainder of his property and for disturbance if he has to move his house or business. The landowner may agree with the Highway Authority for the payment of part of this compensation in kind, for example by way of construction of a private bridge or underpass across the new highway to join two parts of a farm severed by the line of a new road. Alternatively, he may be able to compel the Authority to buy the remainder of his land under section 8 of the Compulsory Purchase Act 1965, unless the Authority can show that part of the property can be taken without material detriment to the remainder or, in the case of the garden of a house, without serious effect on the amenity of the house.

If the owner of a house lost even a small part of his garden to a highway scheme, he could claim compensation for the value not only of the piece of land, which might be relatively small, but also for injurious affection to the rest of his property. This might amount to a much larger sum to compensate him for the depreciation in the value of his house because, from being a house situated in quiet countryside, it might have become a house close to a heavily trafficked new road. On the other hand, until 1973, owners of properties which were physically untouched by the new road would receive no compensation, however close the road passed to their homes or businesses. The Land Compensation Act 1973 sections 1 to 19 has made a major change in the principles underlying compensation law.

Now, if owners or occupiers of property close to a new road can show that the value of their interest in the property, whether as an owner or lessee, has been substantially reduced by the noise, smell, fumes, smoke, vibration or artificial lighting arising from the use of the road, they can claim compensation for that depreciation. Compensation is also available where an existing road is improved to take more traffic, perhaps by widening it or dualling it. In order to allow time for the impact of the new or improved road to be ascertained fairly, claims for compensation have to be made twelve months after the works have been completed and the road, or the improved road, has been opened to traffic.

In the event of the Highway Authority and the owner being unable to agree the compensation payable for a compulsory acquisition, or under the 1973 Act, for the depreciation in value of nearby property, the matter can be referred by either party to the Lands Tribunal for the Tribunal to determine the amount of compensation payable.

The procedures for compulsory acquisition of land for highways and for other public purposes, such as schools and hospitals, are governed by general legislation beyond the scope of this book. The principal Act is the Acquisition of Land Act 1981. Likewise the procedures for assessing compensation are governed by other general Acts and in particular the Land Compensation Acts 1961 and 1973.

Chapter 8

# Making Up Private Streets

Roads and streets are not always highways. Even if they are highways, they are not necessarily maintainable at the public expense. 'Road' is not defined in the Highways Act 1980 and 'street' is defined as having the same meaning as in Part III of the New Roads and Street Works Act 1991. This provides that a 'street' means:

" ... the whole or any part of any of the following, irrespective of whether it is a thoroughfare:–

(a) any highway, road, lane, footway, alley or passage,

(b) any square or court, and

(c) any land laid out as a way whether it is for the time being formed as a way or not."

The definition, therefore, covers both highways and other roads.

The following highways are maintainable at the public expense:

– Any highway in existence before 31st August 1835.

– Streets or roads which become highways after that date where it can be demonstrated that, at some time after dedication as a highway, they have been maintained at the public expense.

– Highways constructed by the Secretary of State for Transport or a local Highway Authority.

– Highways constructed by local councils and Housing Action Trusts under Housing Act powers where construction has been carried out to the satisfaction of the local Highway Authority.

– A highway built and dedicated under an agreement entered into between the builder and the Highway Authority under section 38

of the 1980 Act or previously section 40 of the Highways Act
1959.

## Maintenance of unadopted roads

That may seem a fairly extensive list but there are still hundreds,
and probably thousands, of roads which have been built since 1835
for which responsibility has never been accepted by the Highway
Authority. Another simpler way of putting this is to say that there
are a large number of unadopted roads in the country. Many of these
roads will have become highways because no steps have ever been
taken to prevent members of the general public passing and
repassing over them, but they are still not maintainable at the public
expense. Whilst they may be called private roads, the 'private' refers
to the roads being privately maintainable and is not conclusive as
to whether the road may not, over the passage of time, have become
dedicated as a non-maintainable highway for the use of the public.
Many of these unmade roads were laid down between the turn of
the century and 1960. As the population increased, standards of
living and of housing improved and more and more new housing
estates were developed. Since the enactment of section 40 of the
Highways Act 1959, and more recently section 38 in the 1980 Act,
for the making of Road Agreements between developers and
Highway Authorities, the modern practice has been for builders to
lay out their roads to specifications approved by the local Highway
Authority in advance so that when the roads have been completed
they will be adopted. Consequently, it is not uncommon, as towns
and villages have expanded, to find a new maintainable highway
laid out on a modern estate connected via an older, unadopted,
unmade highway to a main road which is the old maintainable
highway running through the centre of the town or village.

Individual owners of properties close to an unadopted highway may
be liable to repair it by reason of their ownership of the adjoining
or nearby lands. This is known as 'liability ratione tenurae'
(translated roughly as liability by reason of ownership). Such a
liability can be established by proving that for a number of years
owners of those lands have repaired the road in question but it has
been argued that this liability has to be shown to have existed from

time immemorial, before 1189. If it can be shown that the road was laid out later, liability can be rebutted. In practice, it is usually not possible to fix a duty to repair on the owners of properties fronting most unadopted roads (the frontagers). Even where it can be shown that there is a duty to repair laid on adjoining owners, it is still an undecided question whether the persons responsible for carrying out repairs can be sued for damages by a person who sustains personal injuries through failure to repair.[1] The cases are conflicting.

Sometimes, where there is a Residents' Association, private roads, whether they are highways or not, are kept in a good state of repair by the Association. There is a useful power for the Highway Authority under section 296 of the 1980 Act to carry out works on unadopted roads by agreement with, and at the expense of, the persons entitled to execute repairs – for example, the frontagers. A Residents' Association can, therefore, arrange and pay for the Authority to surface their roads from time to time. Usually membership and subscriptions to such an Association are voluntary. Sometimes there may be covenants imposed on the sale and resale of each property under an estate scheme for the purchasers of the houses to contribute towards the maintenance of the roads.

Under section 230 of the 1980 Act, "where repairs are needed to obviate danger to traffic in a private street" the Highway Authority can step in and may, by notice, require the owners of the premises fronting the street to execute, within a limited time, such repairs as may be specified. In the event of failure to execute such works, the Authority can carry out the repairs and recover the costs from the frontagers. A person who is aggrieved by a notice to carry out repairs can appeal to a Magistrates' Court.

If, during the time specified for carrying out the repairs, the majority in number or rateable value[2] of the owners of premises in the street

---

1 *Rundle v. Hearle* 1898.

2 Every proprty has a rateable value. Local taxes payable to the Council and the Water Authority are assessed by reference to the rateable value of the property and known as rates. Since 1990, the Community Charge (Poll Tax) has replaced rates on private homes (though not businesses). In 1993, the Community Charge may be replaced by Council Tax. So far, no amendments have been made to the Highway Acts to replace references to rateable value.

so require, the Authority can be compelled to make up the street under the Private Street Works Code at the expense of the frontagers and, on the completion of the necessary works, the street will become a highway maintainable at the public expense.

### The Private Street Works Code

Section 230 is really an emergency power. The Highway Authority's principal means for having private streets brought up to standard is to operate the provisions of the Private Street Works Code. This has effect for securing execution of street works anywhere in England and Wales. The definition of 'street' has already been given at the beginning of this chapter. Making up of private streets is dealt with in Part XI of the 1980 Act from sections 203 to 237 and begins with an interpretation section of its own. This provides that:

> "In this part of this Act private street means a street that
> is not a highway maintainable at the public expense."

This definition covers both an unadopted highway and a private road which has not been dedicated as a highway. It has been held in a number of decided cases that it is irrelevant whether the street has been dedicated as a highway or not.[1] Frontagers who wish to resist a resolution to make up their street must therefore show that the road is not a street within the statutory definition.

The statutory definition of 'street' makes no reference to the number of buildings on either or both sides of the street. In ordinary, everyday use a 'street' is understood to mean a road in a town or village and not a road in the middle of the countryside. At common law, it has been laid down in decided cases that 'street' means a way which is not necessarily a highway and which has on one or both sides a more or less continuous and regular row of houses.[2] This common law definition is not mentioned in the Highways Acts, and it has been held by the Courts that the statutory definition is wider than that.

---

1  *West End Lawn Tennis Club v. Harrow Corporation* 1965; *Warwickshire County Council v. Adkins* 1967.
2  *Cowan v. Hendon B.C.* 1939.

In *Warwickshire County Council v. Atherstone Common Right Proprietors* (1964) the Divisional Court had to consider a Private Street Works Scheme relating to an extension of Westwood Road, Atherstone. The first part of Westwood Road was a maintainable highway and the County Council wished to make up the rest of the road. The second part of the road was 560 feet long with a railway running along most of one side. Slightly over half the frontage of the other side was taken up by playing fields for 299 feet, there were four dwellinghouses fronting 201 feet, vacant land fronting 39 feet and a small opening for a public footpath over the playing fields. Westwood Road ended in a cul-de-sac with a gate leading to a factory and four cottages and nowhere else.

The Justices were of the opinion that the road was not a street and quashed the apportionment of expenses on the frontagers. Whilst the Divisional Court appreciated that the statutory definition of a street did not require houses on either side of the way and could "give rise to absurdity and oppression if it were pushed to its ultimate literal limits" they considered that, on the facts of this case, the whole of Westwood Road was a street. The Judge giving the decision of the Court said:

> "This is a case where the road is an extension of a public repairable highway; it is a road which already has a number of houses fronting upon it; it is a road which serves a number of premises and must in its turn serve a great many members of the public who make it their business to visit one or other of those premises."

The Court therefore decided that the Private Street Works Code could be applied to the second part of Westwood Road. Theoretically the Code could be applied to a road running between fields but the Court in this case pointed out that there had been no record of any oppression or absurd use of the Private Street Works powers by any local authority in such a way since the powers were first enacted in the late 19th century.

## First and Second Resolutions

Section 205 of the 1980 Act provides that:

"Where a private street is not, to the satisfaction of the Street Works Authority, sewered, levelled, paved, metalled, flagged, channelled, made good and lighted, the Authority may from time to time resolve with respect to the street to execute street works and subject to the Private Street Works Code, the expenses incurred by the Authority in executing those works shall be apportioned between the premises fronting the street."

"Street Works Authority" is, outside London, the County Council or the Metropolitan District Council and inside London is the relevant London Borough. It is, therefore, synonymous with the Highway Authority.

Under section 206 the Authority can include the laying of foul sewers to provide main drainage to properties in the street as well as the laying of surface water sewers to drain the street.

Once the Authority has passed the First Resolution to execute the street works, their surveyor or engineer will prepare a detailed specification of the works to be carried out, an estimate of the probable expenses of the works and a provisional apportionment apportioning the estimated expenses between the premises in the street liable to be charged. The Authority then considers the specification, the estimate and the provisional apportionment and, if they wish, approves them in a Second Resolution. Once the resolution of approval has been passed, notice has to be given in two successive weeks in the local press, posted in the street and served on each of the owners of premises shown in the provisional apportionment as likely to be charged. The notice served on each owner has to be accompanied by a statement of the sum apportioned on his property by the provisional apportionment. The notice must explain that the Authority has resolved to carry out street works in the street, set out where the specification, estimate and provisional apportionment can be inspected and also set out that any owner liable to be charged has the right to object to the provisional apportionment within a period of one month.

## Apportionment of expenses

Under section 207, the apportionment of expenses is to be made "according to the frontage of the respective premises" but the Authority may "if they think just" resolve that in settling the apportionment, regard shall be had to two other matters. These are the greater or less degree of benefit to be derived by any premises from the street works and the amount and value of any work already done by the owners and occupiers of any premises to improve the street. In addition, the Authority may, if they think just, include in the apportionment any premises which do not front the street but have access to it through a court, passage or otherwise and which will, in the opinion of the Authority, be benefitted by the works and they can fix, by reference to the degree of benefit to be derived by those premises, the amount to be apportioned on them. An Authority has complete discretion as to whether or not to use one or more of these criteria in making the apportionment.

Many Authorities will fight shy of having regard to these other matters which are of a subjective nature and can raise more objections, or at least more objections which have a better chance of success before the local Magistrates, than an apportionment based on the frontage of the respective premises, which is a clear objective criterion. On the other hand, it is not entirely equitable that owners with long frontages should pay a great deal more than owners with short frontages, even though both households may each have one or two cars and make exactly the same amount of use of the road. By the same token, the residents in the private street will be keen that the Authority uses its power to include other premises reached via their street in order to reduce the amount of the apportionment falling on the premises in the street. If, however, those other premises are on a modern road constructed and adopted under a Section 38 Agreement, the residents there may consider that the price of their houses included an element of the cost of making up their road and will not wish to pay anything towards making up the older road even though they have to travel along it to get to the main road. It is an area fraught with problems for street works authorities and where it is almost impossible to be fair to everyone.

Another problem is that under section 203(3) 'fronting' includes

115

'adjoining' and under section 329 'adjoining' includes 'abutting on'. Therefore, premises with flank frontages on the private street are liable to be charged, even if they have no access onto that street. If the street is a cul-de-sac and there is a field or house at the end of the street, again with no access onto the street, that part can still be included in the apportionment. Indeed, all the premises "liable to be charged" have to be shown in the provisional apportionment even if the street works authority is not intending to charge those premises or is intending to make a reduced charge. This is to enable all the other owners to see how each property is dealt with and to assist them in deciding whether or not to object with regard to the amount apportioned to their premises. Under section 236, the street works authority does have power to bear the whole or a portion of the expenses of any works in a private street. Without prejudice to this general power, sub-section (2) of section 236 goes on to provide that the Authority may at any time resolve to bear the whole or a portion of the expenses which would otherwise be apportioned on the owner of any premises of which the rear or a flank fronts the street.

A maisonette or flat on the upper floors of a house does not front or adjoin the street and cannot be included in the apportionment unless the Authority resolve to include premises which do not front the street but have access to it.[1]

Local authorities have powers to prescribe a frontage line for erecting new buildings in a street under section 74 of the 1980 Act or through the imposition of conditions on planning permissions for new buildings. In compliance with local authority requirements builders will therefore set new houses and other new buildings back from the carriageway and may leave an intervening strip of several feet between the boundaries of the new properties and the edge of the carriageway. They may not necessary dedicate this strip as a highway. In *Warwickshire County Council v. Adkins* (1967) the Divisional Court considered this sort of situation. The County Council had resolved to make up a road which was roughly surfaced over sixteen feet of its width. Builders developing new houses on

---

1 *Buckinghamshire C.C. v. Trigg* 1963.

the southern side of the road a few years previously had been required to leave twelve feet of land between the edge of the road and the new houses undeveloped. They had not sold this land to the individual householders but had allowed each householder a right of way over it to reach the sixteen foot carriageway. The householders objected to the apportionment on the ground that their premises did not front the street. The Court held the twelve foot strip of land had become part of the private street, that it was irrelevant whether or not it had been dedicated as a highway, and that the expenses of making up the street had been correctly apportioned on the frontages.

Although all the premises have to be shown in the provisional apportionment, churches are not liable for Private Street Works charges and the Authority has to bear their proportion of the expenses. Railway undertakers and canal undertakers are also exempted from street works charges if their land has no direct communication with the street and if their land was solely used for a railway or canal when the street was laid out. In their case, the expenses which would otherwise have been apportioned on their premises are shared between the other premises in the street unless the Authority resolve, under section 236, to bear them as well.

## Objections determined by the local Magistrates

Within one month of the first publication in the local press of the notice that the street is to be made up, any owner of premises shown in a provisional apportionment as liable to be charged with part of the expenses can object to the Authority on the grounds that:

(a) the alleged private street is not a private street;

(b) there is a defect in the resolution to make up or in the specifications and estimates;

(c) the works are insufficient or unreasonable;

(d) the estimated expenses are excessive;

(e) any premises ought to be excluded or included in the provisional apportionment;

(f) the provisional apportionment is incorrect in respect of some matter of fact or in regard to degree of benefit to be derived by any premises or the amount or value of any work already done to improve the street where the Authority have resolved to take these matters into account.

If objections are received and the Authority still wish to proceed, they must refer the objections to the local Magistrates for them to determine what should be done. Objectors must be notified of the hearing by the Magistrates and have the right to be heard. The Magistrates can quash, in whole or part, the specification, plans, estimates and provisional apportionment.

Not surprisingly, not all frontagers respond with enthusiasm to the notice of provisional apportionment and will try to oppose the scheme on one or other of the grounds provided in the legislation. They may, for instance, be able to argue successfully that the street was a highway before 1835 and that, therefore, the costs of making it up must be borne by the Authority.[1] Also, there appears to have been a loophole in the wording in the Highway Act 1835. Under section 23, any highway dedicated after 1835 would only become repairable by the inhabitants at large if notice was given of intended dedication and if the road was made in a substantial manner. However, the section referred to roads, occupation ways and horsepaths. The Divisional Court held in *Richmond (Surrey) Corporation v. Robinson* (1955) that footpaths were not included in section 23. Accordingly, the Private Street Works Code could not be applied to footpaths, even though they might only have been dedicated to the public after 1835. This case has been distinguished in subsequent cases but not overruled. For instance, in *Margate Corporation v. Roach* (1960) it was held that if the land in question was in fact laid out as a road or occupation way, the provisions of section 23 of the 1835 Act would be applicable even though the only public use made of it was by persons on foot.

Even if it is found that a public footpath does lie between private premises and the private street and that, therefore, strictly speaking, the premises do not front the street, the premises can be charged

---

1  *Huyton with Roby U.D.C. v. Hunter* 1955; *Roberts v. Webster* 1967.

with the cost of making up the private street. In *Ware U.D.C. v. Gaunt* (1960) the Hertfordshire County Council tried to resist an apportionment which included the Ware Grammar School, maintained by them, on the basis that a public footpath lay between the school and the private street. The Court held that their premises, though not fronting on the street, could properly be said to adjoin it and that street works charges could, therefore, be apportioned on the school site.

The wording of the First Resolution to make up the street is important and must be followed. If it does not mention lighting, the Authority cannot get round this by approving a specification which includes street lighting when they pass their Second Resolution, nor can the Magistrates include lighting if it is not included in the First Resolution. Similarly, if the Resolution refers to 'channelled' but makes no reference to sewering then the Authority cannot lay drainage pipes under the street. Also, if the Authority intends to exercise its discretion to take degree of benefit into account, it should include in its Second Resolution the words that they think this just.[1]

The Authority can only use degree of benefit to apportion premises which do not front the street but have access to it via a court, passage or otherwise. This means that they cannot make up the first half of the street but include in the apportionment charges against properties lying further along the same street even though those properties can only be reached by travelling over the portion to be made up.[2]

The Magistrates cannot require the Authority to exercise its discretion to include any or all of the three optional criteria: value of works already done, relative degree of benefit to be derived by any premises from the street works and inclusion of properties which do not front the street but which will be benefitted by the works. Equally, they cannot require the Authority to make any contribution to the cost of the works under section 236, or to

---

1 *Oakley v. Merthyr Tydfil Corporation* 1922.

2 *Chatterton v. Glanford Rural District Council* 1915.

increase the amount of contribution which they have decided to make.

Objections on the grounds that the works are insufficient or unreasonable or that the estimated expenses are excessive can be less clear cut and may require Magistrates to hear evidence on engineering detail. In *Bognor Regis U.D.C. v. Boldero* (1962) frontagers objected that works were unreasonable on two grounds. First, they objected to a provision of the Authority's scheme under which the foul sewer would be laid first and then the surface water sewer laid two or three years later to allow the earth to consolidate. Secondly, they argued that it was unreasonable to confine the works to be done to the two streets in question since there were four other streets on the estate which required making up and it would be cheaper if all the roads on the estate were dealt with at the same time. On the first point, the Divisional Court upheld the decision of the local Magistrates to quash the specification after the Magistrates had heard expert engineering evidence that two sewers could be laid at the same time without problems of settlement arising. On the second point, they held that the Authority could not be forced to make up more than the two streets at one time and that it would not have been proper to quash the specification because other streets were not included.

In *Southgate Borough Council v. Park Estates (Southgate) Limited* (1954) the Court of Appeal held that the Justices had jurisdiction to quash a scheme because Park Estates had obtained planning permission to develop land which they owned further up the road and it would have been a waste of money to pay for the first part of the road to be made up if it would be destroyed by construction traffic. In the Bognor case however, the Court said that the Southgate case was a case which depended on its very special facts. The Courts will normally be slow to get heavily involved in engineering details and the Lord Chief Justice said in the Bognor case:

> "The Justices, in considering whether the works are reasonable or unreasonable or sufficient or insufficient, are limited to the reasonableness of the works to carry out the original scheme or object, as set out in the

original Resolution. Their powers in no way extend to
the criticism of that original Resolution. That is a matter
of policy for the local authority."

If a footway is to be made up on one side of the road only, the cost
still has to be apportioned to the properties on both sides of the road.
In *Clacton Local Board v. Young and Sons* (1894) the Local Board
resolved to make up Marine Parade including paving and kerbing
a footpath on the northern side of the road. This street had buildings
on the north side only and was bounded on the south side by land
owned by the Board. It was a grassed area between the road and
seashore. The Court held that the Board were wrong to apportion
the costs of making up the footway against the properties on the
north side only and to exclude that element of cost from the amount
apportioned to the Board's property on the south side of the Parade.

In *Hornchurch U.D.C. v. Allen* (1938) the Court had to consider an
apportionment in respect of a private street called Betterton Road
which was a connecting road between an arterial road leading from
Tilbury to London and a road leading from Dagenham to
Hornchurch. It was agreed that the road would be used by through
traffic. Essex Quarter Sessions allowed an objection that the works
were unreasonable on the basis that a twenty foot carriageway
would have been sufficient for the frontagers to Betterton Road and
that the Authority should not have resolved to specify a twenty-four
foot carriageway in order to cope with the through traffic. The
Divisional Court held that Quarter Sessions were wrong and that it
was within the powers of the Authority to specify a road which
would be sufficient for through traffic as well as the local traffic.
They recognised, as had been recognised in previous cases, that this
might be thought to throw an unfair burden on the frontagers but
considered that the wording of the statute was quite clear and that
this was the intention of Parliament. As explained above, the
Authority could have made a contribution towards the cost but
could not be compelled to do so by the Court.

What then can the Justices or an Appellate Court do on considering
objections to a provisional apportionment? In *Chatham
Corporation v. Wright* (1929) the Magistrates had found that the
works were not unreasonable and that the expenses were not

excessive, but they considered that the apportionments on two of the frontagers were too high and should each be reduced by £100. The Divisional Court held that they could not do this and Mr. Justice Avory said:

> "Unless the Justices intended or contemplated by their decision that the local authority thereby would be impelled to increase their contributions by the sum of £200, the whole scheme became impracticable: and if they either intended or contemplated that, then, as my Lord has pointed out, they were doing something which they had no right to do as appears from the decision of this Court in *Chester Corporation v. Briggs*. If the local authority cannot be compelled indirectly to increase their contribution, it appears to me to follow that the Justices must do one of three things on a further consideration of this case. If they are still of the opinion that the apportionments on the premises of these two owners ought to be reduced, they must consider whether the apportionments on the premises of the other frontagers ought to be increased so as to satisfy the balance of the £200. If they are still of the opinion that the apportionments on the premises of the frontagers ought not to be increased, it appears to me that their only proper course would be to quash the whole of the apportionment, subject to this, that it is open to them to adjourn the consideration of the matter in order to give the local authority an opportunity of considering whether they will, under section 15 (of the Act then in force) pass a new Resolution increasing their contribution by the sum of £200. It appears to me that those are the three courses that are open to the Justices, and they must adopt one or other of them."

If the Justices quash an apportionment, the authority has complete discretion as to whether to prepare a fresh scheme and specification or abandon their intention to make up the street. They cannot be forced to prepare a fresh scheme. If they do, they will again have to give notice to the frontagers of the scheme and of the amount apportioned on each frontager's property, there will again be a period for objections, and if there are objections, these have to be referred to the Magistrates to be heard anew.

## Final apportionment

When the works have been completed and the actual expenses are known, the Authority has to make a final apportionment, dividing the expenses in the same proportions as those in which the estimated expenses were divided in the original provisional apportionment, or the apportionment as amended by the Magistrates, as the case may be. Notice of the final apportionment has then to be served on the owners of all the premises affected and within one month they can again object but the grounds are more limited. They are:

(a) That there has been an unreasonable departure from the specification and plans.

(b) That the actual expenses have "without sufficient reason" exceeded the estimated expenses by more than 15%.

Those objections too, if any, have to be referred to the Magistrates.

Under section 212, the sum apportioned on any premises by the final apportionment, or, as the case may be, that apportionment as amended by local Magistrates, together with interest, is a charge on the premises and is registrable as a local land charge so that future purchasers are aware that there are outstanding road works charges. The Authority has discretion to declare that the expenses apportioned on any premises by the final apportionment can be payable by annual instalments within a period not exceeding thirty years, together with interest. Although the Authority does not have to allow payment over a period by instalments, the heavy costs of making up roads are more than most frontagers can easily find in one sum, particularly retired people who cannot raise further money on their mortgages. It is, therefore, usual for local authorities to allow a period of several years for repayment. This means that they will have to find the initial capital outlay out of their own capital funds or borrow in order to carry out the private street works. In the last decade, restrictions on public expenditure, and in particular on Local Authorities' spending and borrowing, have greatly reduced the number of private street works schemes carried out. As a result, there are still a very large number of unadopted roads in the country, many of them in a bad state of repair.

Frontagers who are aggrieved by the charges made under the final apportionment do have one last chance to pursue the matter, even if the Magistrates have approved the final apportionment. They can, under section 233, appeal to the Secretary of State for Transport, who may make such decision as to him seems 'equitable'. This is a much wider power than the carefully specified grounds of objection to the provisional and final apportionments. For instance, the Secretary of State will sometimes use his power to reduce the charge made on a particular property and require the Authority to bear the amount of the reduction as part of its contribution to the street works whereas the Magistrates have no power to do this.

## The Advance Payments Code

Under sections 219 to 225 of the 1980 Act an Advance Payments Code applies to private streets in London, and in all areas of counties where it was in force immediately before the 1st April 1974. County Councils as Highway Authorities can also adopt the Code in parishes in which it has not yet been adopted. The Code offers an extremely useful mechanism for ensuring that new roads are built to the standards required for adoption by the local Highway Authority. It is, therefore, better than using the Private Street Works Code which brings the road up to adoption standards after it has been built and then only in accordance with the fairly cumbersome procedure outlined above.

The Advance Payments Code catches the building operations at the outset by establishing a procedure which comes into operation as soon as plans for new buildings are deposited in accordance with the Building Regulations, as has to be done with every new building. Section 219 of the 1980 Act provides that no work is to be done for the purpose of erecting the new building unless the landowner has paid to the street works authority or secured to the satisfaction of that authority, the payment to them of such sum as may be required in respect of the cost of street works. If work commences on constructing a building before advance payment has been made or security given, then the owner of the land or, if he is a different person, the builder is guilty of an offence for which he can be convicted and fined.

The Advance Payments Code does not apply if the building to be erected is situated in the grounds of an existing building, if the street is not likely to be joined to a highway maintainable at the public expense, if most of the street is already built up, if the street will serve mainly industrial premises or if there is a Section 38 Agreement providing for the carrying out of works to build the road to adoption standards.

Within six weeks of the passing of the plans, the Authority must serve a notice on the person who submitted the plans requiring payment of such sum as, in their opinion, would be recoverable under the Private Street Works Code in respect of the frontage of the proposed building, or security for such sum. If a builder is laying out a completely new street, and owns all the land on which the street and the houses fronting the street will be built, he will be required to deposit an amount equivalent to the total of all the charges which would be made on the frontages, i.e. the cost of making up the whole street. There is a right of appeal against the notice to the Secretary of State if the person served considers that the sum required is excessive. If the builder, as is often the case, proceeds to lay out the road himself, then the Authority can refund part of the money deposited or release the security to the extent of the work done as the work proceeds. When the work has been completed satisfactorily to adoption standards, the whole sum paid will be released or the security will be discharged. Similarly, if a Section 38 Agreement[1] is entered into after notices have been served under the Advance Payments Code and monies deposited, the monies will be released as work proceeds.

Section 222 provides that where a sum has been paid or secured under the Advance Payments Code, the liability of the frontage owner or any subsequent owner in respect of the carrying out of street works under the Private Street Works Code will, as respects

---

1  See Chapter 3.

that frontage, be deemed to be discharged to the extent of the sum so paid or secured.

All sums paid by the owner of the land to the Authority under the Advance Payments Code carry simple interest from the date of payment until such time as the sum falls to be set off against charges for street works carried out by the Authority or until it falls to be refunded because the builder has completed the road to adoption standards.

Notices requiring payment under the Advance Payments Code, and payments made, and securities given, are registrable as local land charges. Charges for street works are heavy and no purchaser wants to spend a large sum buying a house and then to find that he has to spend further monies paying such charges. The standard practice is, therefore, for a purchaser's solicitor and his building society solicitor to make enquiries as to whether the road outside the house being purchased is maintainable at the public expense and if so whether there are still outstanding street works charges in respect of works carried out under the Private Street Works Code. If the road is not maintainable at the public expense, the solicitor will want to know whether monies have been deposited or secured to cover street works charges or whether there is a Section 38 Agreement in force for the making up and adoption of the road.

### Adoption of private streets

Where an Agreement is entered into under section 38 of the 1980 Act, the road will become a highway maintainable at the public expense as soon as the works specified in the Agreement have been completed to the satisfaction of the Highway Authority. Where a road has been made up under the Private Street Works Code, or the Advance Payments Code, and the road was not already a highway when the resolution to make it up was passed, adoption as a highway maintainable at the public expense is not automatic. The street works authority may, if they wish to adopt the road upon completion of the works, initiate the procedure laid down in section 228 of the 1980 Act. This requires them to display notices in the street declaring the street to be a highway maintainable at the public expense. The owner of the street or, if more than one, the majority in number of the owners of the street, have a month in which to object to the notices and, if they object, the Authority has a further two months in which to apply to a Magistrates' Court for an order overruling the objection. No criteria are laid down in section 228 to guide the Magistrates in how to determine the objection. It is likely that they will give consideration to the utility of the road to the general public and will have regard to whether it is a through road or not.

The Highway Authority does not have to put in hand the procedure for the street to become maintainable at the public expense. For their part, if street works have been completed to the satisfaction of the Authority, the majority of the owners of premises in the street in terms of rateable value may require the Authority to accept responsibility for future maintenance. In that event the Authority must display the appropriate notices in the street declaring the street to be a highway maintainable at public expense; there will be no right of objection by other owners in the street in this circumstance.

It may be thought that there would be little benefit in the residents objecting to a proposal that their road should become a highway maintainable at the public expense. However, it must be remembered that private streets are not necessarily highways when the street works are carried out. Although the street works authority can require a street to be made up to adoption standards, they cannot require the owners to dedicate it as a highway except under the section 228 procedure. If the residents want to keep the general public out of the road, they may decide that it is in their best interest that the road remain private. One advantage to the residents is that they can put ramps down in the road to slow traffic because the road is not a highway.[1] The disadvantage to them is that the Highway Authority will not maintain the road in the future and, if residents allow it to deteriorate, it is quite possible for the Highway Authority to repeat the street works procedure and again resolve, under section

---

1 Although the Highway Authority cannot interfere with the ramps, residents need to be aware that, if there is an accident, there could be claims against them under the Occupiers Liability Acts in negligence.

205, that the road should be made good to adoption standards at the expense of the frontagers.[1]

In practice, once the road has been made up properly, if it is a quiet residential road, it will last a long time without a great deal of maintenance work being necessary. It is not uncommon to see under street names, in properly surfaced residential roads, words in small print on the lines of "This is a private road. Notice is hereby given that there is no public right of way over the road. By order of the Owners."

If at least one payment has been made by an owner under the Advance Payments Code, and the majority of owners on a built up private street request the street works authority to exercise their powers under the Private Street Works Code to make up the street and to declare it to be maintainable at the public expense, then under section 229 the street works authority must comply unless they release the money which has been deposited by one or more owners.

---

1 *The Barry and Cadoxton Local Board v. Parry* 1895.

# Chapter 9

# Works in the Highway

## The New Roads and Street Works Act 1991

Roads have been dedicated by individuals or constructed by Highway Authorities as highways over which the Queen's subjects have an unchallengeable right of passage. Any interference with that right of passage by acts of obstruction, such as breaking open the surface of the highway, is likely to amount to a public nuisance at common law, be actionable as negligence or fall foul of the numerous penal provisions in Part IX of the 1980 Act, described in Chapter 5. Clear statutory authority is required to enable individuals or companies to break open the highway and lay and maintain apparatus under it. This authority has been given by Parliament to the companies providing gas, electricity, sewerage and telecommunications under Acts such as the Telecommunications Act 1984 and the Water Act 1989.

These are recent Acts dealing with services which have been privatised by the Government in the last few years. These companies, like their nationalised predecessors, still need the power to lay services in the highway in order to reach homes and businesses throughout the country. The Public Utilities Street Works Act was enacted in 1950 in order to set out a code of uniform procedures for the protection of Highway Authorities and the bodies and persons having powers to lay apparatus in the highway. The Street Works Code was designed to cover three relationships:

1. To give some control to the Highway Authority over the way in which apparatus such as pipes and cables are laid and repaired in the highway.

2. To give the Undertakers with responsibility for providing these services protection when the Highway Authority themselves

F

carry out work in the highway, such as reconstituting the surface of the highway or widening or diverting it.

3. To provide some protection as between different Undertakers so that an Undertaker, say, exercising a right to lay a new pipe in the highway, does not damage apparatus already there and belonging to another Undertaker.

The 1950 Street Works Code has been used for over forty years to reconcile the sometimes competing claims between Highway Authorities and Statutory Undertakers. It was considered by some to be weighted in favour of the Undertakers. It has now been replaced by Part III of the New Roads and Street Works Act 1991, sections 48 to 106. Whilst many of the provisions are similar to those in the 1950 Code, there are some significant changes. In particular, Undertakers can be penalised for undue delay in completing works, which will be good news to motorists, they now have to compensate Highway Authorities for damage arising from their works, whether or not there is negligence, and they have to pay fees to the Highway Authority for inspections of their works.

### Terminology in relation to street works

It is necessary to be familiar with the terminology in the 1991 Act in order to understand the provisions made by Parliament for street works. A 'street' means the whole or any part of any highway, road, lane, footway, alley or passage and any square or court, whether or not there is a through way. 'Street' also includes any land laid out as a way whether for the time being it is formed as a way or not. Where the street passes over a bridge or through a tunnel, references in the Act to the street include the bridge or tunnel. It will be appreciated that this definition, contained in section 48(1) of the Act, is much wider than a highway maintainable at the public expense. Indeed, it is expressly stated that the Act applies to a street which is not a maintainable highway subject only to such exceptions as may be prescribed in Regulations made by the Secretary of State for Transport.

Street works means the placing of apparatus and inspecting, maintaining, repairing, renewing and changing the position of the apparatus. It also includes works required to carry out these operations, notably breaking up the street to lay the apparatus and tunnelling or boring under the street.

Under section 48 'Undertaker' includes both persons and companies who have statutory rights to place apparatus in the street and also those to whom a licence is granted by the street authority under section 50. 'Street authority' in relation to the street means, if the street is a maintainable highway, the Highway Authority, and, if the street is not a maintainable highway, the 'street managers'. The phrase 'street managers' used in relation to a street which is not a maintainable highway, means the authority, body or person liable to maintain the street or, if there is none, any authority, body or person having the management or control of the street. Therefore, in private streets this could mean the frontagers on either side of the street if, as is usual, they own the street up to its centre line from the boundary of their property.

It is also important to note that in the scheme laid down for street works in the Act there are other relevant authorities as well as the street authority. These authorities are:

1. Where the works affect a public sewer, the sewer authority – likely to be one of the privatised Water Companies now, such as Thames Water or Severn Trent.

2. Where the street is carried or crossed by a bridge vested in a Railway or Canal Company, that Company.

3. Where, in any other case, the street is carried or crossed by a bridge not vested in the street authority, then the authority in which that bridge is vested – for example, a privately owned bridge where tolls may still be payable.

### Giving notice of street works and the response of the Authority

There are plenty of provisions in the Highways Act 1980 to prevent interference with streets. In addition, section 51 of the 1991 Act makes it an offence for any person other than the street authority to place apparatus in a street or to break up or open a street unless they have either a statutory right or have been granted a Street Works Licence by the street authority. The prohibition does not apply to 'emergency works'. Such works are defined in section 52 as any works required to put an end to circumstances "which are likely to cause danger to persons or property". In the event of a person being charged with carrying out unauthorised street works, the onus is on him to prove that the works were emergency works.

Any Undertaker proposing to execute street works involving breaking up or opening the street, or tunnelling under it, must give not less than seven working days' notice under section 55 to the street authority,[1] to any other relevant authority and to any other person having apparatus in the street which is likely to be affected by the works. A seven day notice given in accordance with these requirements ceases to have effect if the works to which it relates are not substantially begun before the end of the period of seven working days from the starting date specified in the notice. This saves the time and expense of the street authority and any other relevant authority which might otherwise have to keep visiting the site to which the notice relates to see whether works have started in order to supervise them.

Under section 56 of the Act, if it appears to the street authority on receiving the seven day notice that the proposed works are likely to cause serious disruption to traffic, and that the disruption could be avoided or reduced if the works were carried out only at certain times, the authority may give the Undertaker directions as to the times when the works may or may not be carried out. An Undertaker who executes works in contravention of such directions commits an offence. This is a useful new power for the street authority which did not exist in the 1950 Act. It means that they can now have a say in trying to minimise the disruption to traffic except in the case of emergency works.

### Minimising disruption and co-ordinating street works

Furthermore, in the interests of minimising disruption, section 58

---

1  Seven days' notice is the general rule. Section 55 also enables the Secretary of State to provide different periods of notice for different descriptions of works.

empowers the Secretary of State to make Regulations prescribing what are "substantial road works". Where the street authority propose to carry out substantial works themselves for highway purposes, or where they are aware that a sewer authority or a transport authority in whom a bridge over or under a highway is vested, or one or other of the Undertakers, intend to carry out substantial road works, they may restrict the execution of street works during the twelve months following the completion of the substantial road works. The start of the substantial road works must be at least three months away and a notice of the proposal to carry out the works has to be given to the sewer authority, to any bridge authority where the part of the highway, to which the restriction will relate, is carried or crossed by a bridge and to all the other Undertakers having apparatus in the part of the highway to which the restriction relates. The purpose of these provisions is to enable every other body with apparatus in that length of street to carry out any necessary repair work or alterations to their apparatus during the three month notice period, or while the substantial road works are in progress, so that the street is not dug up two or three times within a relatively short period. Co-ordination of street works in this way should also lead to financial savings since, if one Undertaker breaks open the street, other Undertakers can reach their apparatus without having to carry out that work themselves and reinstatement can be carried out on behalf of them all.

If the street authority impose a restriction under section 58, no Undertaker can break up or open the part of the highway to which the restriction relates during the following twelve months except to execute emergency works. If he contravenes the restriction he commits an offence punishable on conviction by fine of up to £1,000. These fines may not seem very large for big companies like British Gas or British Telecom but under the 1950 Code there were no duties on Undertakers to co-operate, no powers to give directions that work should only be carried out at certain times (section 56) or that no work should be carried out for a period of twelve months and consequently no financial penalties at all for failing to co-operate or causing the street to be dug up several times in quick succession.

The 1991 Act also introduces, under section 59, for the first time the general duty on the part of the street authority to co-ordinate street works and a general duty on Undertakers, under section 60, to co-operate. These duties are imposed:

1. In the interests of safety.

2. To minimise the inconvenience to persons using the street (having regard in particular to the needs of people with a disability).

3. To protect the structure of the street and the integrity of apparatus in it.

The street authority must "use their best endeavours to co-ordinate the execution of works of all kinds" including works for road purposes and must have regard to any Code of Practice issued by the Secretary of State giving practical guidance as to how co-ordination can best be achieved. The Codes of Practice can also give practical guidance to Undertakers on how best to co-operate. A failure on the part of an Undertaker to comply with the Code of Practice is evidence of failure to comply with the duty laid on Undertakers under section 60 to co-operate and the Act goes so far as to make an Undertaker who fails to comply with this duty to co-operate guilty of an offence and liable to a fine on conviction of up to £1,000 (section 60(3)).

The general arrangement in both the 1950 and 1991 Acts is that Undertakers give notice to the street authority that they intend to carry out works in pursuance of their statutory powers. They do not need the consent of the street authority but, as explained above, there are now some controls on the way they can work. Disruption will inevitably be caused by the works but the Highway Authority, motorists and other users of the highway have to put up with this. It has however been necessary under section 20 of the Highways Act 1980 for the Undertakers to obtain specific consent from the Highway Authority to place apparatus or lay services in any special road: a special road includes all motorways and this is why it is not usual to see men and vehicles from a Water Company or British Telecom or any other Undertaker working on a motorway.

## Stricter requirements on Undertakers in certain streets

Sections 61 to 64 of the 1991 Act introduced three further categories of street where the Highway Authority will be able to exercise a greater degree of control. These are protected streets, streets with special engineering difficulties and traffic sensitive streets.

In protected streets specific consent will be required from the Highway Authority to placing apparatus or laying services. Section 20 of the Highways Act 1980 has been repealed and all special roads will now be protected streets. In addition, however, the Highway Authority will be able to designate other streets as protected in accordance with criteria laid by the Secretary of State.

Secondly, streets can also be designated as having "special engineering difficulties" and, in such streets, the Undertaker as well as giving notice to the street authority, must in accordance with the procedure laid down in Schedule 4 of the 1991 Act, agree a plan and section of the proposed works with the street authority, and with the sewer authority and bridge authority if the works will affect a sewer or bridge. In default of agreement, there are provisions for arbitration. Streets carried by bridges over railways or canals are likely to be those with special engineering difficulties and a transport authority can ask the local Highway Authority to designate a street as having special engineering difficulties.

The Secretary of State can also make Regulations enabling Authorities to designate a street as 'traffic-sensitive'. Such Regulations may include a special provision with regard to giving advance notice, the starting dates for works and special arrangements for carrying out emergency works which otherwise an Undertaker could begin without giving any notice.

## Safety measures

There is a strong emphasis on taking proper safety measures in carrying out street works and section 65 requires the Undertaker to guard and light any opening in the street or any plant or materials on the street. He is also required to place and maintain traffic signs for the guidance of persons using the street and to have particular regard to people with a disability. Again, if an Undertaker fails to

135

comply with any Code of Practice issued by the Secretary of State, he commits an offence. New provisions have been inserted in section 67 requiring Undertakers to ensure that their street works are supervised by a person having a prescribed qualification as a supervisor and to ensure that at all times when any works are in progress there is at least one person on the site having a prescribed qualification as a trained operative. A fine of up to £1,000 on conviction can also be imposed on vandals removing warning signs and lights or safety fencing.

The common law duty to take care to avoid accidents applies to Undertakers, as well as the statutory obligations on them outlined above to take proper safety precautions. An illustration of the standard of care expected of an Undertaker is given by the case of *Pitman v. Southern Electricity Board* (1978). The Electricity Board had been carrying out some works in Mere in Wiltshire and had dug a trench across the road and a hole in the pavement for the insertion of a junction box. When they reached the pavement at the end of the day's work they had to leave a hole in it because the junction box had still to be inserted and connected up with other cables. They covered the hole with a large metal plate two feet ten inches square and one-eighth of an inch thick. The plaintiff, who was an elderly lady, was walking along the pavement at dusk and tripped over the metal plate. She sued the Board for damages.

In their defence the Board cited the Liverpool pavement cases, described in Chapter 4. Lord Justice Lawton said "in my judgment, that series of cases has little bearing on this case. We are not concerned with the unevenness of flagstones at all. Had this plaintiff tripped over a flagstone with a tiny difference in the level, the result might have been very different at first instance and also in this Court. What she tripped over, as I have already stressed, was the unexpected condition and level of the pavement." All three Judges agreed in holding the Board liable at common law in negligence.

### Completing street works quickly

Section 66 provides that an Undertaker carrying out street works "shall carry on and complete the works with all such dispatch as is reasonably practicable" and that if he fails to do this he commits an offence. Where the Undertaker creates an obstruction in a street to a greater extent or for a longer period than is reasonably necessary, the street authority can require him to take steps to mitigate or discontinue the obstruction; if he fails to take these steps, the authority may put them in hand themselves and recover the cost of doing so from the Undertaker. Section 74 empowers the Secretary of State to provide by regulations that an Undertaker shall be liable to pay a charge to the Highway Authority where the works are not completed within a reasonable period. The charges to be made and the type of street works to which this section will apply will be laid down in Regulations made by the Secretary of State. In the event of dispute about what is a reasonable period, the matter can be referred to arbitration.

## Reinstatement

Sections 70 to 73 impose duties on Undertakers to re-instate the street after they have completed their works. Reinstatement is to be begun as soon as is reasonably practicable and to be completed "with all such dispatch as is reasonably practicable". The Act distinguishes between an interim reinstatement and a permanent reinstatement, which must be completed within six months from the date on which the interim reinstatement is completed. The Authority is to be informed as soon as the reinstatement has been completed so that it can carry out an inspection and, if necessary, investigatory works in order to ascertain whether the reinstatement has been properly executed. Regulations will prescribe the materials to be used and the standards of workmanship to be observed and there may also be Codes of Practice. Where an Undertaker fails to comply with his duties under the Act and Regulations with regard to reinstatement, the Authority can serve notice on him to carry out remedial works. If he fails to comply, the Authority can carry out the works and recover the costs from the Undertaker. He may also be charged with offences under sections 70 and 71.

Section 73 lays down a rough and ready rule as to the responsibility for reinstatements where the road is subsequently dug up by another Undertaker, as can quite often happen. If a reinstatement does not conform to the performance and standards prescribed by

Regulations, it shall be presumed, unless the contrary is proved, that this was caused by the last Undertaker carrying out street works at that site. This should avoid Undertakers blaming each other for faulty reinstatements and make the task of the street authority much easier in affixing responsibility for putting matters right.

### Diversion of traffic

Under section 14 of the Road Traffic Regulation Act 1984, the traffic authority, which is usually the same authority as the street authority and the Highway Authority, can make Traffic Orders diverting traffic onto alternative routes in order to facilitate the execution of street works by an Undertaker. If an order is made for this purpose, the traffic authority can recover from the Undertaker the administrative costs of making and advertising the order and of providing traffic signs indicating the alternative route(s) to motorists. Where the traffic is diverted onto an alternative route of a lower classification, for example from a trunk road to a principal road, or from a classified road to an unclassified road, the Undertaker has to indemnify the Highway Authority for the alternative route in respect of costs incurred by it in strengthening the highway so far as this is necessary for the purposes of its use by diverted traffic, or in making good any damage to the highway occurring in consequence of its use by the diverted traffic (sections 76 and 77 of the 1991 Act).

### Maintenance of apparatus

An Undertaker having apparatus in the street must ensure under section 81 of the 1991 Act that it is maintained to the reasonable satisfaction of the street authority having regard to the safety and convenience of persons using the street and in particular to the needs of people with a disability. The apparatus must also be maintained to the reasonable satisfaction of the sewer authority, and the bridge authority if there is a bridge affected by the apparatus.

### Roads under construction

Section 87 contains arrangements for applying the provisions of the Act with regard to the placing of any apparatus or the laying of services in a street which is likely to become a maintainable highway. The section will be of importance with regard to agreements, under section 38 of the Highways Act 1980, where builders are laying out roads on new estates which they intend to dedicate to the public and to have adopted by the Highway Authority. The Highway Authority can make a declaration that the street is likely to become a maintainable highway. The declaration must be registered as a local land charge. The provisions of the Act with regard to the execution of street works then apply to such a street. The Authority is required to secure the performance by Undertakers of their duties under the Act in such a manner as is reasonably required for the protection of the street managers, who will usually be the builders, until the street is adopted. Thus, if a new residential estate is being developed, and services such as sewers, gas pipes, telephone and electric cables are being laid under it, the work will be carried out in an orderly fashion within the scheme of the Act and there will be proper records of all the apparatus placed in the road before it becomes a maintainable highway.

## Works carried out by Highway Authorities

So far as Highway Authorities themselves are concerned, they, in their turn, must, under section 83 of the 1991 Act, give notice to all the Undertakers with apparatus in the street where they are carrying out road works, must give them facilities for monitoring the execution of the works and must comply with requirements made by the Undertakers which are reasonably necessary for the protection of their apparatus or for preserving access to it. An Authority which fails to comply with these provisions commits an offence.

Where the Authority are carrying out major highway works such as the reconstruction or widening of the highway, the construction of dual carriageways and roundabouts, or even the construction of road humps and cattle grids, they must work closely with the Undertaker whose apparatus is likely to be affected. Under section 84 they must identify any measures needed to be taken in relation to the apparatus, they must settle a specification of the necessary measures and determine by whom they are to be taken and they must co-ordinate the taking of those measures and the execution of their own works "so as to secure the efficient implementation of the necessary work and the avoidance of unnecessary delay".

The Act also lays down duties with regard to registers and records. The street authority must keep a register showing such information as may be prescribed with respect to street works executed or proposed to be executed in the street. This register is to be open for public inspection. Undertakers must keep records of the location of every item of apparatus belonging to them and must make those records available for inspection by the street authority and by any other Undertaker having authority to execute works in the street.

### Street Works Licences

Most apparatus in the streets and most work involving the breaking up or opening of streets is the responsibility of one or other of the major statutory Undertakers, such as British Gas or British Telecom. These companies have powers in their enabling statutes to execute works in the street and, as explained in this chapter, in many cases simply need to give notice to the street authority that they are carrying out works under their statutory powers. The authority cannot veto their work. Other bodies and individuals who want to carry out any work in a street have to apply for a Street Works Licence, under section 50 of the 1991 Act, to the street authority. This section, together with Schedule 3, sets out detailed provisions with regard to licences and replaces sections 181 to 183 of the 1980 Act. It is under section 50 that a person would have to apply for a licence to carry out works, such as laying a private sewer or boring a tunnel under the street, in order to connect up different parts of his property which are separated by the street. Whether or not to grant a licence is at the discretion of the authority. Conditions can be imposed on the licence with regard to safety precautions and minimising inconvenience to persons using the street. There is a right of appeal to the Secretary of State against the refusal of a licence in respect of an application to place apparatus on a line crossing the street but no right of appeal if the apparatus is to be placed along the line of the street.

A Street Works Licence can be granted to a person on terms permitting or prohibiting its assignment or to the owner of land and his successors in title. If the licence is granted to the owner and successors in title, every owner, before transferring the property, must give notice to the street authority stating to whom the benefit of the licence is to be transferred.

Before granting a Street Works Licence, the street authority must consult any Undertaker having apparatus in the street likely to be affected by the licensee's works. The Authority can require the payment of a reasonable fee in respect of their administrative expenses in granting the licence and an annual fee for administering the licence. Conditions can be attached to the licence in the interest of safety and to minimise the inconvenience to persons using the street.

Although a licence can be granted for the benefit of an owner and successors in title, every licence can be withdrawn by the authority on three months' notice if the authority consider it necessary for the purpose of the exercise of their functions as a street authority. If a licensee fails to comply with the Act or any condition of the licence, the licence can be terminated on seven days' notice.

## Compensation for damage or disruption caused by street works

Whether or not Undertakers should compensate street authorities when matters go wrong and the authority suffers damage or loss through the apparatus being in the street has been a vexed question, with some cases going as far as the House of Lords, such as *Department of Transport v. North West Water Authority* (1983). In that case, the House of Lords had to consider the provisions of the 1950 Act and decided that North West Water was carrying out a statutory duty to provide water via pipes under a trunk road and that, when those pipes burst through no fault of the water authority, they were not liable to compensate the Minister for the ensuing damage to the highway.

Section 82 of the 1991 Act now provides that an Undertaker shall compensate the street authority, any other relevant authority and any other person having apparatus in the street in respect of any damage or loss as a result of the execution of street works by the Undertaker or as the result of any explosion, ignition, discharge or other event occurring to gas, electricity or water or any other service in apparatus belonging to the Undertaker in the road. Section 82(3) provides that the liability of an Undertaker under this section arises whether or not the damage or loss is attributable to negligence on his part, and notwithstanding that he is acting in pursuance of a statutory duty.

The compensation will be payable even if the Undertaker is executing a statutory duty and is not negligent. The same strict liability is applied to licensees under a Street Works Licence who have to indemnify the street authority against any damage or loss arising out of the placing in the street of their apparatus, whether or not they are negligent.

These provisions relate to liabilities between an Undertaker and the street authority and between different Undertakers. The Act is silent as to liability on the part of persons executing street works to third parties such as adjoining owners. Section 95 simply says that any provision imposing criminal liability in respect of any matter is without prejudice to any civil liability in respect of the same matter. As a general rule bodies exercising a statutory power will not be liable for damage caused through the exercise of that power unless they have exercised it negligently. In *Lagan Navigation Co. v. Lambeg Bleaching, Dyeing and Finishing Co. Ltd.* (1927) Lord Atkinson said:

> "If a man or public body have statutory powers which he or they may exercise in a manner hurtful to third parties or in a manner innocuous to third parties, that man or body will be held guilty of negligence if he chooses or they choose the former method in exercising his or their powers and not the latter, both being available."

This principle was followed by the Court in *Levine v. Morris* (1969) discussed in Chapter 4 where the Department of Transport was held liable for the negligent siting of a road sign into which a vehicle

crashed because the sign could have been ere~
whilst serving the same purpose.

On the other hand some authorising statutes impose on ~
a liability to pay compensation whether or not there .
negligence, in the same way as section 82 discussed above impo~
such a liability as between the Undertakers and the Highway
Authority.

This is obviously of importance to persons whose property may be
damaged by, say, the bursting of a water main under a street or
whose business may be adversely affected by the time Undertakers
take to carry out works outside their premises to the inconvenience
of their customers, resulting in loss of trade. *Leonidis v. Thames
Water Authority* (1979) was a case where the owner of a motor
repair garage was able to recover compensation for loss of trade
while the street outside his premises was closed to enable the water
authority to reconstruct a sewer. The authority had the duty to do
this work under the Public Health Act 1936 and were not negligent
but that Act provides for compensation to be paid for any loss caused
through the execution of any duty under the Act.[1]

So far as the Highway Authority is concerned, they have wide
powers to carry out works in the highway, they can authorise
Undertakers to carry out street works and they can divert traffic by
means of Traffic Orders. Unless they exercise these powers
negligently they are not normally liable for injury to persons or
damage to property occurring through the execution of works in the
highway. Whilst there is a general provision in the Public Health
Act 1936 providing for compensation for anyone suffering loss
through the exercise of powers under that Act, there is no such
straightforward provision in the 1980 Highways Act, the Road
Traffic Regulation Act 1984 or in the more recent 1991 Act dealing
with street works. There appear to be no recent cases where a
Highway Authority has been held liable for loss of business to shops
caused through traffic diversions and there is an old case[2] where a

---

1  Public Health Act 1936 section 278. See now Schedule 12 Water Industry Act 1991.

2  *Martin v. L.C.C.* 1899.

greengrocer failed in such a claim. One of the Court of Appeal Judges said:

> ".... I am strongly inclined to think that the damage alleged to have been suffered by the plaintiff, which is damage of the same nature and kind as everyone else in the street suffered through the road being blocked, is not of such a character as to give a cause of action."

With regard to Statutory Undertakers, the stronger powers available to the Highway Authority under the 1991 Act to give directions under section 56 as to when works may be carried out, and under section 66 to take steps to put an end to obstructions caused by Undertakers taking longer than is reasonably necessary for their work, could be a double-edged sword. Although these are powers rather than duties, a Highway Authority might be open to actions for judicial review or to complaints to the Local Government Ombudsman if it does not give proper consideration to exercising those powers in appropriate cases.

The 1991 Act is very recent, Part III only came into force on 1st January 1993 and not all the detailed Regulations and Codes of Practice on which so many of its provisions depend have yet been issued by the Secretary of State. It is early days to see how the new arrangements between Authorities and Undertakers will affect the structure of streets, the standards of reinstatement and the users of the highway.

Chapter 10

# Traffic Regulation

The importance of the highway as a facility open to all subjects of the Crown and the right of every citizen to pass along the highway has, it is hoped, been apparent in every chapter of this book. The origins of the special status of highways reach back to the Middle Ages, and even earlier for the Roman roads. Until relatively recently, there have been few restrictions on the way in which people can use highways.

Members of the public and occupiers of property fronting the highway must not use the road in such a manner as to cause a public nuisance by obstructing it, damaging it or interfering with it in some other way. Users of the highway have a duty of care towards other users and will be liable in negligence for injuries they cause to others through careless driving in the event of an accident occurring. There is no right to use vehicles on a bridleway or footpath. However, this would be no more than an act of trespass in respect of which only the landowner could bring an action, unless the vehicular use damaged the surface of the way and made it impassable for pedestrians or horseriders, in which case an action might be taken in public nuisance.

The relative lack of restrictions on the way in which highways could be used worked well enough in the days of horses and carts. Until 1936, drivers did not even have to take a test before driving a car on the road. With the enormous increase in the volume of traffic using the highways, and in the speed and power of vehicles on the road, Parliament has found it necessary in the second half of the twentieth century to impose a vast body of statute law on the way in which the public use highways. If they had not done so, it would not be possible to control excessive speeds or drunken driving; nor would it have been easy under the limited scope of actions in public nuisance to restrict the use of narrow lanes by heavy vehicles, nor to reduce congestion and increase safety in town centres by the use of one-way streets and pedestrianisation of shopping centres.

There are now two major legislative codes dealing with these matters. The Road Traffic Acts deal with who can drive on the roads and try to secure competent and safe driving and roadworthy vehicles. The Road Traffic Regulation Acts make it possible for Highway Authorities to regulate how the roads are to be used, for example by making orders about the speed and direction of traffic, and provision for parking places in some roads and prohibitions on parking in others.

## The Road Traffic Act 1988

The Road Traffic Act 1988 (the 1988 Act), as amended by the Road Traffic Act 1991, makes provision for the major driving offences, such as causing death by dangerous driving, careless and inconsiderate driving and driving whilst under the influence of drink or drugs. The 1988 Act also deals with the licensing of drivers of vehicles, and tests of their competence to drive and the wearing of seat belts by motorists and crash helmets by motor-cyclists. It makes separate provision for the licensing of drivers of heavy goods vehicles. Part II of the Act contains extensive provisions about the construction and use of vehicles, such as the arrangements for test certificates for cars more than three years old. Part VI covers compulsory insurance against third party risks.

## The construction and maintenance of vehicles and their use

The safe manufacture of vehicles is largely secured by type approval requirements, either as to vehicle parts such as steering, brakes and tyres, or as to a vehicle as a whole, imposed under section 54 of the 1988 Act. Nearly all new vehicles require a type approval certificate or similar certificate. Under section 63 of the 1988 Act it is an offence to use on a road, or cause or permit to be used on a road, a vehicle or vehicle part in respect of which a type approval certificate is required unless that requirement is complied with. The relevant requirements are largely related to European Council Directives or to Regulations annexed to the Agreement about the adoption of uniform conditions of approval for motor vehicle

146

equipment and parts concluded at Geneva in 1958 ('ECE Regulations'). The current type approval requirements are mostly contained in the Motor Vehicles (Type Approval) (Great Britain) Regulations 1984 and the Motor Vehicles (Type Approval for Goods Vehicles) (Great Britain) Regulations 1982. Both these sets of Regulations have been subject to numerous amendments. The safe manufacture of vehicles and vehicle parts is also partly secured by Regulations under section 41 of the 1988 Act, mainly the Road Vehicles Lighting Regulations 1989 and the Road Vehicles (Construction and Use) Regulations 1986. Some of the requirements of these Regulations are related to European Council Directives or ECE Regulations and there is an overlap between type approval requirements and the requirements of Regulations under section 41.

## Regulations prohibiting unnecessary obstruction

Most of this body of Regulations is concerned with ensuring that vehicles are in a safe condition before they are used on the road, but it is worth noting that Regulation 103 of the Road Vehicles (Construction and Use) Regulations 1986 provides that "no person in charge of a motor vehicle or trailer may cause or permit the vehicle or trailer to stand on a road so as to cause any unnecessary obstruction of the road".

What is an "unnecessary obstruction"? This is a difficult issue. Attention was drawn in Chapter 5 to a number of cases, not all of them going the same way, in which the Courts have considered prosecutions for obstruction. Highways are for passing and repassing. It is not the case that the absence of waiting restrictions and yellow lines means that there is a right to park vehicles in the road. There is no such right. The Judges have tried to take a practical view and to treat parking in case of accident or for rest and refreshment as permissible.[1] When dealing with prosecutions under the Regulation for causing an "unnecessary obstruction" the Courts have tended to regard the matter as one of degree. In *Solomon v. Durbridge* (1956) the Divisional Court considered that leaving a

---

1  *Rodgers v. M.O.T.* 1952.

vehicle in the road for five and a quarter hours amounted to an unnecessary obstruction.

In that case, Mr. Solomon, a Barrister, had left his car from 11.00 a.m. to 4.15 p.m. parked on the Victoria Embankment, London, where there were no restrictions on parking, while he worked in his Chambers in the Temple. The City of London Magistrate who heard the case found that there was an unnecessary obstruction of the road and convicted Mr. Solomon. The Divisional Court upheld the conviction. The Lord Chief Justice said:

> "I entirely fail to see how it can be said that a stationary car is not an obstruction on the highway; it certainly is, because it obstructs the free passage. One cannot walk over or drive over a place where a stationary car is standing, but such are the exigencies of modern life that no-one is going to say that leaving a car for a reasonable time is an obstruction sufficient to make it an offence. Therefore, the Regulations in question in this case have been careful to make it an offence, not merely to obstruct, but to cause unnecessary obstruction."

**Parking and driving off the road**

Sections 19A, 21, 22 and 34 of the 1988 Act place some restrictions on parking and driving off the road. Under section 19A, a person must not park a vehicle on the footway or verge of a road which is subject to a thirty mile an hour or forty mile an hour speed limit, or on any land situated between two carriageways of a road subject to such a speed limit. Contravention of this section is a fixed penalty offence. Under section 21, no person may drive a motor vehicle on a cycle track. Under section 22:

> "If a person in charge of a vehicle causes or permits the vehicle or a trailer drawn by it to remain at rest on a road in such a position or in such condition, or in such circumstances as to be likely to cause danger to other persons using the road, he is guilty of an offence."

This provision is more specific than the offence of unnecessary obstruction in the Construction and Use Regulations, discussed above. It catches motorists who park their cars close to junctions or bends.

Under section 34, it is an offence for a person without lawful authority to drive a motor vehicle on any footpath or bridleway or upon any common land or moorland. The phrase "without lawful authority" means that the landowner can drive a vehicle along a footpath across his own land so long as he does not damage the surface. Under subsection (2) of section 34, it is not an offence to drive a motor vehicle on any land within fifteen yards of the road "for the purpose only of parking the vehicle on that land". Section 34 is a useful means of preventing motorcycling along footpaths or across open land used for recreational purposes by the general public.

The Refuse Disposal (Amenity) Act 1978 also contains useful powers for the local council to remove any vehicle which appears to have been abandoned on the highway, or anywhere else in the open air.

**The Highway Code**

No law says that drivers of vehicles should travel on the left-hand side of the road. This is stated in the Highway Code but, until the passing of the Road Traffic Act 1930, there was not even a Highway Code. Section 38 of the Road Traffic Act 1988 is the current statutory provision governing the Highway Code and provides that the Secretary of State may from time to time revise the Highway Code in such manner as he thinks fit. It is not an offence to break the Highway Code, but section 38(7) provides that:

> "A failure on the part of a person to observe a provision of the Highway Code shall not of itself render that person liable to criminal proceedings of any kind, but any such failure may in any proceedings (whether civil or criminal, and including proceedings for an offence under the Traffic Acts, the Public Passenger and Vehicles Act 1981, or sections 18 to 23 of the Transport Act 1985) be relied upon by any party to the proceedings as tending

to establish or negative any liability which is in question in those proceedings."

Paragraph 43 of the present Code instructs drivers "keep to the left". Not to drive on the left would not in itself be an offence but in view of section 38(7) quoted above, and the universally understood convention in this country that vehicles are driven on the left, failure to do so would make a prosecution under section 2 of the 1988 Act for dangerous driving virtually certain to succeed.

## Traffic Regulation Orders

Under section 1 of The Road Traffic Regulation Act 1984 (the 1984 Act) Traffic Regulation Orders may be made:

(a) for avoiding danger to persons or other traffic using the road;

(b) for preventing damage to the road or any building near the road;

(c) for facilitating the passage on the road of any class of traffic (including pedestrians);

(d) for preventing the use of the road by vehicular traffic of a kind which, or its use by vehicle traffic in a manner which, is unsuitable having regard to the existing character of the road or adjoining property;

(e) for preserving the character of the road in a case where it is especially suitable for use by persons on horseback or on foot;

(f) for preserving or improving the amenities of the area through which the road runs.

The Authority for making orders is the Secretary of State for Transport in respect of motorways and trunk roads. For other roads in London, the London Borough is the appropriate Authority. Outside London, the County Councils and the Metropolitan District Councils are the traffic authorities in the same way as they are the Highway Authorities.

Under section 2 a Traffic Regulation Order can be made to achieve one or more of the above objects by:

Restricting the use of the road to particular types of vehicle.

Prohibiting vehicular traffic from using the road.

Prohibiting the use of the road by through traffic.

Prohibiting the use of the road by pedestrians.

Making a road into a one-way street.

Prohibiting or restricting the waiting of vehicles.

Prohibiting or restricting the loading and unloading of vehicles.

Prohibiting the use of the road by heavy commercial vehicles.

The Authority may also include in a Traffic Regulation Order, provision to specify the through routes for heavy commercial vehicles or for restricting the use of heavy commercial vehicles, in such zones as may be specified, as they consider expedient for preserving or improving the amenities in their area and subject to such exemptions as they think fit.

Under section 4 an order imposing restrictions may also contain exemptions and arrangements for the grant and display of certificates for vehicles which are exempted, such as disabled persons badges or residents' parking permits.

These are very wide powers, particularly when it is appreciated that most Traffic Regulation Orders do not need the consent of a Secretary of State or Minister. The Secretary of State has power to make Regulations laying down the procedure to be followed in making orders and the current Regulations were made in 1989.[1] The effect of these Regulations is that, except in special cases, the Authority simply has to advertise the draft order it intends to make in the London Gazette and in the local press and to allow a period of three weeks for objections and representations to be submitted in response to the advertisement. The Authority must then consider any objections received and, having done so, can resolve to make the order in the form originally intended and advertised.

---

1   S.I. 1989 No. 1120.

Alternatively, if they choose, they can modify the order or abandon it. One example of the extent of the order-making power is that, by prohibiting the use of a road by vehicular traffic, an Authority can reduce its status in practice to that of a footpath without having to go through all the rigours of a Stopping Up Order, described in the final chapter of this book.

A Traffic Regulation Order may also, by directing traffic away from certain streets, have a detrimental effect on shops in those streets, particularly those which depend upon passing trade. Similarly, restrictions on parking on a highway in areas close to shops often call forth the wrath of the shopkeepers, who fear that most of their customers do not want to walk more than ten yards to reach them. This may be less so nowadays as the attractions of pedestrianisation to shoppers have come to be appreciated. An Authority, in making a Traffic Regulation Order, is required, under section 122 of the 1984 Act, to do no more than have regard to the desirability of securing and maintaining reasonable access to premises. A shopkeeper cannot claim compensation for loss of trade as a result of traffic being directed elsewhere.[1]

Whilst a local authority can make orders prohibiting the use of any road by through traffic, they do need the consent of the Secretary of State to make the order if the order will have the effect of preventing for more than eight hours in any period of twenty-four hours access for vehicles of any class to any premises situated on or adjacent to that road. Ministerial consent is also required to an order imposing a thirty mile per hour speed limit on a principal road, or an order imposing a speed limit of less than thirty miles per hour on any other road.

The procedure for making a permanent order takes several weeks in order to allow time for the order to be advertised for three weeks and for objections to be considered.

### Inquiries

In certain circumstances the Regulations made by the Secretary of

1  *Martin v. L.C.C.* 1899.

State do require the Traffic Authority to hold a Public Inquiry before making an order. These instances arise when a draft order has been advertised prohibiting the loading or unloading of vehicles in any road except during the rush hours and an objection is made to the order. Also if a bus company objects to a draft order prohibiting the use of a street by buses, or requiring traffic to proceed in one direction only along the street, there will have to be an inquiry. The Traffic Authority must select an inspector to hold the inquiry from a panel of persons chosen by the Secretary of State to hold inquiries of this nature and must consider his report and recommendations before making the order.

Whilst most orders can be made by the Highway Authority without the consent of the Secretary of State and without a local inquiry, the validity of an order can be challenged in the High Court on the grounds that it is not within the statutory powers of the Authority or that any of the relevant requirements, such as advertising, have not been complied with. As will be seen later in this chapter, there have been challenges in the High Court to some orders.

## Temporary orders

There are speedier procedures available in cases where works are being executed in the road, possibly by a Statutory Undertaker such as a gas company, or where there is a likelihood of danger to the public, or serious damage to the road. Under section 14 of the 1984 Act,[1] an Authority may, by order, restrict or prohibit temporarily the use of the road. When considering the question of the making of such an order, they should have regard to the existence of alternative routes suitable for the traffic which will be affected by the order. Such a temporary order has to be advertised for seven days in the local press. It does not have to be advertised in the London Gazette.

In general a temporary order will not continue in force for more than eighteen months. The Secretary of State can give consent for

---

1 Section 14 has been rewritten by the Road Traffic (Temporary Restrictions) Act 1991. The periods for the duration of the orders are those set out in section 14 as substituted by the 1991 Act.

a further six months. Temporary orders on footpaths and bridleways must not last longer than six months.

Under section 14(2), there is an even speedier provision which can be brought into force without delay, for example in the event of serious flooding or a burst gas main. In such a case the Authority may, at any time by notice, restrict or prohibit temporarily the use of a road but such a notice shall not continue in force for more than five days from the date of the notice or twenty-one days in certain circumstances.

Temporary orders and notices can also be used to impose speed limits when roads are being surface dressed in order to compel motorists to drive slowly and avoid throwing up chippings into other motorists' windscreens.

Temporary orders or notices must not prevent access for pedestrians to any premises situated on or adjacent to the road affected.

It is also possible, under a useful extension to section 14 added by the Environmental Protection Act 1990, to make an order or issue a notice closing a road temporarily, or restricting parking, in order to arrange for clearing litter or street cleaning. This is a helpful power to enable the District Councils to arrange for a street to be cleaned without problems of through traffic, or parked cars at the kerbside, making it extremely difficult to pull leaves and other debris out of the gutters and manholes.

## Speed limits

Part VI, sections 81 to 91 of the 1984 Act contains powers to impose speed limits. Section 81 imposes a general 30 miles per hour limit on all restricted roads. Restricted roads are roads where there is a system of street lighting provided by means of lamps placed not more than 200 yards apart. The Highway Authority for the road can issue a direction that this automatic limit is not to apply to a specified road even if it has such a system of street lighting. Equally, under section 84, the Highway Authority can impose a 30 mile per hour or 40 mile per hour limit on roads which are not restricted roads.

The Act also contains powers for the Secretary of State to make speed limit orders on a national basis. Regulations have been made restricting speeds on motorways and dual carriageway roads to 70 miles per hour anywhere in the country and on single carriageway roads to 60 miles per hour.

Section 86 of the 1984 Act also imposes speed limits on motor vehicles of particular classes. Thus, for example, articulated vehicles having a maximum laden weight exceeding 7.5 tonnes are restricted to 60 miles per hour on motorways, 50 miles per hour on dual carriageways and 40 miles per hour on any other road.

## Parking places

Under section 1 of the 1984 Act, a Highway Authority can prohibit parking in roads. The Act also contains provisions for local authorities to take positive steps to facilitate the free flow of traffic and to provide proper parking places. Both the County and District Councils can provide off street car parks under section 32 of the Act. These can be surface car parks, underground car parks or multi-storey car parks and in all cases the local authorities can make a charge for their use. In order to impose a charge on an off street car park, the local authority providing it needs to make an order under section 35 and can include in that order regulations enabling them to remove vehicles left in the car park in contravention of the order. It is usual in these orders to provide that vehicles cannot be left overnight in the car park.

Under section 45 of the Act, the County Councils and the District Councils can make orders designating parking places on the highway for on street parking. If a District Council, not being a Highway Authority, makes such an order, it needs the consent of the Highway Authority to the order. The order can make provision for only certain types of vehicles to park in the road and can also include Regulations about the length of time the vehicles can park. The council can charge by using parking meters. Another option open to the council is to restrict on street parking to persons of a particular class who are made eligible to purchase a permit. Residents' parking schemes can be made in this way under section 45.

The positive provisions in sections 32 and 45 of the 1984 Act empowering local authorities to provide parking places on and off the highway are an essential corollary of their regulatory powers to prohibit parking. Indeed, section 122 of the Act lays down that:

> "It shall be the duty of every local authority ... so to exercise the functions conferred on them by this Act as ... to secure the expeditious, convenient and safe movement of vehicular and other traffic (including pedestrians) and *the provision of suitable and adequate parking facilities on and off the highway.*"

### Special occasions, processions, fairs, emergencies

There are three powers under which action can be taken by local authorities or the police to deal with problems of congestion on the highway on special public occasions or in emergencies.

Under section 21 of the Town Police Clauses Act 1847, the District Council can make orders for the route to be observed by vehicles, and for preventing the obstruction of the streets:

> "In all times of public processions, rejoicings, or illuminations and in any case when the streets are thronged or liable to be obstructed..."

No special formalities are required to make this sort of order but it can only last for the duration of the special occasion, such as a fair. An order for a continuous period of more than a day, or at the very most two or three days, would be beyond the powers of the section.[1]

Under section 287 of the Highways Act 1980, either the County Council or the District Council can in any case of an emergency, or on any occasion on which it is likely:

> "By reason of some special attraction that any street will be thronged or obstructed ..."

arrange for barriers to be erected in the street and kept in position for so long as may be necessary for that purpose.

---

1 *Brownsea Haven Properties Ltd. v. Poole Corporation* 1958.

Under section 67 of the Road Traffic Regulation Act 1984, a police officer may place in the highway such traffic signs indicating prohibitions, restrictions or other requirements relating to vehicular traffic as may be necessary or expedient to prevent or mitigate congestion or obstruction of traffic "in consequence of extraordinary circumstances". The power to place signs conferred by this section includes power to maintain them for a period of seven days. Failure by motorists to obey such a sign is an offence under section 36 of the Road Traffic Act 1988.

### Consultation with the police and enforcement of Traffic Orders

It is common sense for local authorities to consult the police before they make orders prohibiting through traffic or heavy commercial vehicles using a particular road, or for prohibiting parking in the road, or for imposing speed limits. If the police do not have enough resources to enforce the orders, it is doubtful whether it is worth making them. In fact, under Part III of Schedule 9 to the 1984 Act, the local authorities are required to consult with the relevant Chief Constable or, in London, with the Metropolitan Commissioner of Police, before making an order for any of these purposes. They do not have to accept the advice they receive from the police but run the risk of the order being ignored by motorists if the police say they will be unable to assign officers to enforce the order.

Orders which can be self-policing do not suffer from problems of enforcement. Under section 92, in the 1984 Act, the Highway Authority is empowered to place bollards or other obstructions at any point in a road where the passage of vehicles is prohibited to traffic altogether (and the premises fronting the road have rear vehicular accesses). Bollards can be placed at either end of the road to stop vehicles travelling down it. So far as trying to keep heavy commercial vehicles out of a road, a width restriction is much easier to enforce than a weight restriction, because bollards can be placed on either side of the road narrowing the width down to allow for the passage of private cars only, whereas with a weight restriction the police need to be present to stop the heavy vehicles using the road.

Under the Road Traffic Offenders Act 1988, breach of certain

Regulation Orders is a fixed penalty offence, for which the penalty is currently twenty pounds. However, it is at the discretion of the police whether they give the offender a fixed penalty notice or not. If they do not, then upon conviction by the Magistrates for contravention of the Traffic Regulation Order, an offender can be fined up to one thousand pounds. Contraventions of Traffic Regulation Orders imposing speed limits are, of course, also penalised by the imposition of penalty points, endorsements and disqualifications, depending upon the previous record of the offender.

## London

The arrangements for making Traffic Regulation Orders are slightly different in London, but the powers to make the orders are also contained in the 1984 Act. Section 6 and Schedule 1 of the Act give the London Boroughs powers to make Traffic Regulation Orders in a rather wider range of circumstances than in the rest of the country. Nevertheless, they can also make orders for exactly the same purposes as those set out in section 1(1) of the Act for Authorities outside London. In addition, Part II of the Road Traffic Act 1991 made further special provision for traffic in London. Under section 50 the Secretary of State can designate priority routes in London.

Under sections 63 to 77 of the 1991 Act, it will be possible to introduce a tougher regime for controlling parking in London. Whilst, under section 65, contravention of orders imposing waiting restrictions will not be a criminal offence, local authorities in London have been given stronger powers to immobilise vehicles, to remove and dispose of vehicles parked contrary to Traffic Regulation Orders and to recover fixed penalties. Under Schedule 6, if a person fails to pay a fixed penalty charge the local authority can serve a Charge Certificate increasing the penalty by 50% and can apply to the County Court for an order enabling them to recover the increased charge as if it were payable under a County Court Order. The operation of these new provisions is being watched with interest because, under section 43 of the 1991 Act, Highway Authorities outside London can apply to the Secretary of State to extend similar powers to them.

## Orders for the preservation of the amenities of an area

Perhaps the most important case in recent years on the extent to which Authorities can go in making Traffic Regulation Orders was the *R. v. the London Boroughs Transport Committee* on the motion of the Freight Transport Association. It arose in London on an order made under section 1(1)(f) and section 6 of the 1984 Act:

> "For preserving or improving the amenities of the area through which the roads runs."

This power is available to traffic authorities outside London as well as inside.

*R. v. The London Boroughs Transport Committee* was decided by the House of Lords in July 1991. The London Boroughs Transport Committee had taken over an order made by the former Greater London Council. That order had prohibited heavy commercial vehicles driving through residential areas at night-time and at weekends except with a permit. Condition 11 attached to the permit required the permit holder to fit air brake silencers and hush kits. The Freight Transport Association, the Road Haulage Association and five transport operators applied for orders of judicial review to quash Condition 11 on the ground that it was beyond the powers of the Road Traffic Regulation Act. The Associations and the other plaintiffs won, both before the Divisional Court and the Court of Appeal. However, they received short shrift from the House of Lords.

Lord Templeman said:

> "Many owners have fitted suppressors, as requested. Others have driven and still drive through residential Greater London without reducing the sound level of their £30,000 vehicles by fitting a £30 suppressor, preferring to leave a nightly trail of wailing babies, disturbed invalids and cursing residents on the principle that 'none shall sleep'."

The Associations submitted to the House of Lords that the London Boroughs were going beyond their powers under the 1984 Act. Lord Templeman continued:

"One policy and one object of the 1984 Act were, however, to protect the environment of Greater London. Condition 11 had been intended and was effected to carry out that policy and fulfil that object."

Judgment was given for the London Boroughs and the condition was upheld.

Attention was drawn earlier in this chapter to the Construction and Use Regulations and an argument had also been advanced in the Divisional Court by the Associations and other applicants that Regulations about braking systems were entirely a matter for the Secretary of State for Transport under the Construction and Use Regulations. It is interesting to note that by the time the case reached the House of Lords, Counsel for the applicants had abandoned this argument and did not even ask the House to rule on it.

### Enforcement of orders

It is an offence under section 5 of the 1984 Act to contravene a Traffic Regulation Order. Under section 68 of the Act the Highway Authority can erect traffic signs of a size, colour and type prescribed by national Regulations made by the Secretary of State to indicate the effect of an order, for example a one way street or a no right turn sign. It is an offence under section 36 of the Road Traffic Act 1988 for a driver to fail to comply with the instructions given on certain signs, in particular the no entry and the one-way signs.

### Usefulness of Traffic Orders

Whether or not to make orders controlling traffic is often a controversial issue. Early examples were not encouraging. In 45 BC, Julius Caesar forbade any wagon to be led or driven during the daytime within the built up area of Rome. In 1635, Charles I issued a Proclamation that the "general and promiscuous use of coaches" in the streets was not only a great disturbance to His Majesty and to others of Place and Degree but also broke up the pavements and made passage for pedestrians more dangerous. He, therefore, commanded that no hackney or hired coach should be used in London. It is doubtful whether the orders of either Julius Caesar or

Charles I were obeyed and the untimely fate of these two men, no doubt at the hands of the motoring interests at the time, has been a grim warning to future generations of traffic engineers.

The powers in the 1984 Act for local authorities to make Traffic Regulation Orders can be relatively easily exercised and are widely used. The Authorities do, however, have to tread a difficult path between opposing interests in the way they regulate the use of roads. Local residents will always press for more speed limits, more residents' parking schemes and prohibitions on anybody else parking in their roads. Heavy vehicle bans are also much in demand, but that can equally make things difficult for local trades and businesses in transporting their goods within the locality. In any case, the shops in every town depend on heavy goods vehicles reaching them and Highway Authorities have to recognise that, even if their residents do not.

From the highway engineer's point of view, the primary purpose of Traffic Regulation Orders is to avoid danger to persons or other traffic using the roads and to ensure the free movement of traffic along the roads. Even in trying to achieve these relatively modest aims, he can expect to encounter difficulties. If there is a dangerous junction between a minor road and a major road and a Traffic Regulation Order is made closing the junction, so that residents living in the minor road have to go to work by driving along to the other end of their road, and then possibly by a longer route to reach the major road, they may object, even though there may have been a very bad accident record at the junction. Shops like to have on-street parking near them and object to waiting restrictions being imposed outside their premises as it reduces their passing trade. On the other hand, unrestricted parking leads to traffic jams developing all the way down the High Street and accidents occurring through pedestrians emerging from behind parked vehicles.

It is not necessary to stop up a highway in order to create a pedestrian precinct, for example in a shopping centre. A Traffic Regulation Order can prohibit the use of the street by vehicles, providing it does not prevent access for vehicles for more than eight hours in the day. Thus, orders can be made for excluding traffic, for example, from 11.00 a.m. to 5.00 p.m. It is also possible to make an exemption for loading and unloading vehicles.

Stopping up a highway is a complicated and lengthy procedure, as will be seen from the final chapter. Co-operation of the Statutory Undertakers is required in any stopping up procedure. This will not be necessary where a Traffic Regulation Order is made because the street remains a highway and the Undertakers still have their powers to lay and maintain apparatus in the street. The use of removable or lockable bollards can also enable emergency vehicles, such as fire engines and ambulances, to have access at all times.

The powers in the Act to be flexible in the times of day and of the year when restrictions are imposed can be enormously helpful. With increasing car ownership, and greater numbers of holidaymakers each year, Cornwall County Council found that traffic conditions in the attractive fishing villages were becoming impossible during the summer. There might only be one narrow road leading down to the sea with little room for turning when the road reaches the seashore. The council were able to make Traffic Regulation Orders forbidding traffic other than residents to drive down the village street during the daytime between April and September and they provided large car parks at the entrance to the villages. The first such orders were made in the village of Polperro and were initially strongly resisted by the shopkeepers. However, it soon became apparent that the absence of vehicles made the street a great deal more pleasant for visitors and encouraged more rather than less tourists. Other villages then began to press for similar orders. This type of seasonal restriction can now be seen in many coastal resorts throughout the country.

In 1978 the Berkshire County Council made an experimental Traffic Order in order to preserve and improve the amenities of their area. Under this order they prohibited lorries over 5 tonnes in weight using short stretches of all the roads on the south side of Windsor in order to reduce the heavy through traffic in the town. The effect of the order was that the lorries had to make a detour of several miles and the Freight Transport Association and the National Farmers' Union applied to the High Court to have the order quashed on the grounds that it was beyond the powers of the council.

The High Court, and then the Court of Appeal,[1] said that the County Council had behaved quite properly and that the order was valid. The order contained provisions for owners of any commercial vehicles to apply for exemption certificates and also for any persons owning businesses in Windsor to apply for exemption certificates for particular vehicles to serve their premises. The Courts decided that the council had therefore fulfilled their duty, by including these provisions for exemptions in the order, to have regard to the desirability of maintaining reasonable access to premises.

Some Authorities have been able to use Traffic Regulation Orders to reduce the number of accesses to industrial estates at night-time. This has been done at the suggestion of the police because thieves feel uneasy about entering an estate to break into premises if there is only one way in, and the same way out, so that their vehicles are more likely to be noticed. Orders have been made closing some of the roads leading into the estates from 9.00 p.m. to 5.00 a.m. The purposes for which orders can be made under section 1 of the Act, as set out earlier in this chapter, do not include the prevention of crime, but it is unlikely that potential offenders would wish to draw attention to themselves by seeking to have such orders quashed. It would probably be going too far to argue that the orders fall within section 1(1)(b) on the ground that they may help to prevent damage "to any building on or near the road".

### The meaning of 'road'

Both the Road Traffic Act 1988 and the Road Traffic Regulation Act 1984, discussed in this chapter, apply to roads which are not necessarily highways, as well as to highways. In the 1984 Act, 'road' means:

> "Any length of highway or of any other road to which the public has access."

In the 1988 Act the wording is almost identical. 'Road' there means:

> "Any highway and any other road to which the public has access."

---

1  *Freight Transport Association v. Berkshire County Council* 1980.

The meaning of "any other road to which the public has access" was considered in the case of *Adams v. The Metropolitan Police Commissioner* (1979). Mr. Adams was a resident in the Aberdeen Park Estate in north London. There was a Residents' Association which owned the roads and levied contributions from the residents to maintain them. Notices were put up by the Residents' Association saying that the roads were private roads and that access was forbidden to unauthorised persons.

Despite these notices, a large number of people living outside the estate used the roads. There was a church, a doctor's surgery, a centre for urban educational studies, a Foreign Missions Club and a hotel. It could be argued that members of the public were coming to these buildings on the estate because they were invitees, rather than as of right. Thus, though the roads had not been dedicated as highways, there was much traffic on them.

In addition, the Judge found that there was a very high degree of pedestrian usage. Many pedestrians used the roads through the estate as a short cut from their homes on one side to shops on the other. Other people came into the estate to walk dogs, and children came in to collect 'conkers' falling from the chestnut trees which lined the roads. On Saturday afternoons, Arsenal supporters parked their cars on the estate roads.

The litigation arose because rather less desirable members of the public also began using the roads. Youngsters sped through the roads in cars and on motor-cycles. The residents asked the police to enforce the Road Traffic Acts and to prosecute drivers for speeding and for reckless driving. The Metropolitan Police Commissioner had been advised that the estate roads were not roads to which the public had access and that, therefore, the Road Traffic Acts were not being contravened. The Judge decided that the public did have access and that the Acts could be enforced on the estate roads.

After the proceedings had been issued, but before the case was heard, the Residents' Association changed the notices at the entrances to the estate to remove the references to access being restricted to residents and their callers, leaving only the statement that the roads were private. The Judge said that, although it had been understandable for the Commissioner to have taken the view that the public had no access to the roads in view of the wording of the original notices, the notices must be looked at as part of the whole picture. The decisive point was whether the public use of the roads was tolerated by the owners. The Judge found that it was, despite the notices, and that the public had access.

The roadways in a car park, whether public or private, may also be 'roads' for the purposes of the 1984 and 1988 Acts. If a Court found that the public had access to them it could convict for careless driving or driving while uninsured.[1]

---

1 *Oxford v. Austin* 1980.

# Chapter 11

# Footpaths, Bridleways and Byways

A footpath is a highway over which the public have a right of way on foot; a bridleway is a highway over which the public have a right of way on foot and on horseback; and a byway open to all traffic ('a byway') is a highway over which the public have a right of way for vehicular traffic but which is used mainly for the purpose of walking or horseriding.[1] Under the Wildlife and Countryside Act 1981, a *public path* means either a footpath or a bridleway and a *right of way* means a public path or a byway open to all traffic. When the phrase 'right of way' is used in this chapter it will, as provided for in the 1981 Act, include footpaths, bridleways and byways.

The three types of way are all highways. The general law of highways applies to them. They have come into existence through dedication or statutory creation, as described in Chapter 2. The Highway Authority has a duty to maintain them as much as if they were metalled roads. A user of the way suffering an accident through failure to maintain has a right of action against the Highway Authority. It is just as much an offence to obstruct a public footpath or to damage its surface as it would be with a road. A right of way cannot be stopped up or diverted, any more than a road which is a highway can be stopped up or diverted, except by going through the strict statutory procedures described in the final chapter of this book. Why then write a chapter about rights of way?

Although footpaths and bridleways were the earliest forms of highway, they are today the most likely routes to disappear off the map unless action is taken to protect them. The legislature has accepted this and, beginning with the Rights of Way Act 1932, there

---

1 Wildlife and Countryside Act 1981 s.66(1).

has been an increasing volume of legislation dealing specificai
with rights of way. It is the purpose of this chapter to set out tha.
legislation. A number of causes led to a loss of rights of way before
Parliament started to take action. The drift of the population from
the countryside to towns following the Industrial Revolution meant
that there were far fewer people to walk the paths in the countryside.
The invention of the motor car resulted in far more journeys to work
being carried out by car or bus and far fewer paths being used every
day as a means of going to and from work. The outward sprawl of
towns and villages and the covering of more green fields each year
with houses has also led to the loss of rights of way. Finally, whereas
it takes many years for a well made road to disappear, and Roman
roads can still be traced today, an unsurfaced footpath can become
overgrown and impassable within one summer, if it is not walked
regularly, and invisible after two or three summers.

At the same time as paths have been lost during this century, through
falling into disuse, or being built over, it has become easier for
townsfolk to reach the countryside by train, bus or car. Shorter
working hours have enabled more people to look for recreational
opportunities in the countryside. The Countryside Commission
estimate that many millions of people regularly walk in the
countryside.

Wealthy owners of large houses and grounds in the country are not
always ready with a warm welcome for walkers and horseriders
crossing their land on rights of way; farmers are understandably
worried about security of their stock and damage to their growing
crops. These conflicting interests have led to a good deal of tension
between different pressure groups, such as the National Farmers'
Union and the Ramblers' Association and to some unpleasant
confrontations on countryside rambles between landowners and
hikers. In the 1930s well-known acts of mass trespass were
organised on the Pennine Moors. The Countryside Commission
claims that a recent survey has revealed that, even on a mere two
mile stroll, walkers have a one-in-three chance of not being able to
complete their journey due to some obstruction and that horseriders
fare even worse, with only a two per cent chance of completing their
course on a typical ten mile ride.

## Rights of Way Act 1932

The Rights of Way Act 1932 enabled the existence of a right of way to be established by showing twenty years' uninterrupted use as of right. It was no longer necessary to prove an act of dedication by a landowner, perhaps many years previously. This Act was discussed in Chapter 2 and its terms are now embodied in section 31 of the Highways Act 1980. It is interesting to read fifty years later what Lord Justice Scott said, with remarkable percipience, in giving his Judgment in the Court of Appeal in one of the early cases on the 1932 Act:[1]

> "If, on such proof as there was here, a footpath cannot be pronounced public, I despair of the future of most of our unregistered public paths, of which there are still very many. By 'unregistered', I mean those not yet officially recorded by the local authority or owners. In these days, when motor buses, motor cars and motorcycles transport so many into the countryside both for business and for pleasure, and when practically all agricultural workers, and indeed most of the rural population, have their bicycles, long footpaths, which fifty years ago meant so much for ease of communication, are infinitely less frequented, and it becomes easier and easier for real public rights of way to disappear, just because they become unproveable. Yet the Rambler – sometimes called 'the hiker' – needs the footpath more than ever. The movement represented by the Ramblers' societies is of national importance, and to the real lover of the country, who knows that to see it properly he must go on foot, but who is driven off all main roads and a good many others by the din and bustle of motor traffic, the footpath is everything. In short, it is of real public moment that no genuine public footpath should be lost, without statutory action to close it."

---

1 *Jones v. Bates* 1938.

## National Parks and Access to the Countryside Act 1949

*Definitive Maps of rights of way*
The Rights of Way Act 1932 was the first step, but the absence of a system of compulsory registration of paths meant that paths continued to be lost. If they were not used, they became quickly overgrown and if their existence was not recorded in a public register somewhere, it was difficult to prove that the path had actually existed on the ground. The next step, therefore, was the imposition by Parliament through the National Parks and Access to the Countryside Act 1949 of a duty on local Highway Authorities to prepare maps showing footpaths, bridleways and roads used as public paths in their areas. For this purpose, the 1949 Act appointed the Highway Authority as the Surveying Authority.

Each Surveying Authority had to consult the District and Parish Councils and assemble information on the existence of the rights of way in their area. They then had to prepare both a draft map on which each right of way claimed was plotted and a statement accompanying the map, describing the right of way. When a draft map had been prepared, notice was to be given by public advertisement and objections could be made by landowners to the inclusion of rights of way on the map. These objections were determined by the Minister of Housing and Local Government and then the Surveying Authority had to proceed to the next step, which was the preparation of a Provisional Map. If there were objections to the inclusion of rights of way on the Provisional Map, those objections had to be determined by Quarter Sessions and, after the Courts Act 1971, by the Crown Court. Once all the objections to the Provisional Map had been dealt with, the Surveying Authority was in a position to publish the Definitive Map of rights of way in their area.

The legislation excluded footpaths at the side of a public road from the Definitive Map.[1]

Footpaths and other rights of way do not have a high priority in a Highway Authority with scarce resources and Authorities could not

---

1 National Parks and Access to the Countryside Act 1949 s.27 and now the 1981 Act s.66.

afford to employ many staff to deal with all the work involved. Since each county has thousands of footpaths, and since objections from landowners were common, it took years for all the paths to be plotted and for the procedures to be finalised.

Most Highway Authorities have now completed their Definitive Maps. The Ordnance Survey have been able to take the information from the Definitive Maps to use in indicating public rights of way on their current 1:50,000 Landranger Series. The keys on these maps, as well as indicating the colouring and symbols used to show the public rights of way, also show all other roads and paths on the ground but point out that the representation on the maps of any other road, track or path is no evidence of the existence of a right of way.

## Long distance paths

The provisions in the National Parks and Access to the Countryside Act 1949 for the preparation of Definitive Maps of rights of way and for keeping those maps up to date have now been repealed and replaced by Part III of the Wildlife and Countryside Act 1981. Before leaving the 1949 Act, however, attention needs to be drawn to two important provisions relating to rights of way in the 1949 Act still in force. Under sections 51 to 55, the Countryside Commission can prepare plans and carry out works to provide long distance routes where it appears to the Commission that the public should be enabled to make extensive journeys on foot or on horseback along a particular route. The Commission's plans will show existing rights of way and proposals for the dedication or compulsory creation of new lengths of path to link up the existing rights of way, so as to form a continuous long distance route. The Highway Authorities remain responsible for these long distance routes, but have been much assisted by grants from the Commission in bringing them into existence. Examples of long distance routes are the South Downs Way and the Pennine Way.

## Notices deterring the public from using rights of way

The other important provision remaining from the 1949 Act is section 57. Under this section it is an offence for any person to place on or near any way shown on a Definitive Map, a notice containing any false or misleading statement likely to deter the public from using the way. Upon conviction for this offence, the Court can make an order requiring the person responsible for the erection or retention of the notice to remove it and may impose a further fine for each day during which the notice is not removed.

Because of the shortage of staff in Highway Authorities to deal with rights of way, there are still a large number of notices in the countryside deterring use of rights of way, such as 'Private Road', or 'No Through Road'. Although these notices do not actually say that there is no right of way, they are likely to put off less knowledgeable members of the public from continuing with a walk past such a notice.

## Countryside Act 1968 – rights of way signs

It was logical for Parliament to take the next step of encouraging the erection of positive signs to indicate rights of way as well as seeking to outlaw the negative signs. Under section 27 of the Countryside Act 1968, it became the duty of the Highway Authority to erect and maintain signposts where a right of way leaves a metalled road, indicating that there is a footpath, bridleway or byway starting at that point and:

> "Showing, so far as the Highway Authority consider convenient and appropriate, where the footpath, bridleway or byway leads, and the distance to any place or places named on the signpost."

The only exception to this rule requiring the beginning of a right of way to be signposted is when the Highway Authority is satisfied that it is not necessary and the Parish Council, in whose area the path is, agrees.

The Highway Authority can also erect and maintain signposts along any right of way, after consultation with the owner or occupier of the land and, under subsection (4):

> "It shall also be the duty of a Highway Authority ... to erect such signposts as may in the opinion of the Highway Authority be required to assist persons

171

unfamiliar with the locality to follow the course of the footpath, bridleway or byway."

Most rights of way are now properly signed at the point where they leave the metalled road, but very few Authorities appear to have paid much attention to subsection (4), quoted above, and waymarking along paths is conspicuous by its absence. Some of the voluntary societies, like the Chiltern Society, do waymark paths in their areas but there is nothing in England and Wales to match the excellent system of waymarking paths in the Austrian and Swiss Alps.

### Riding of pedal cycles

Section 30 of the Countryside Act 1968 conferred an important right on any cyclist, but not on motor-cyclists, to ride a bicycle on any bridleway. In exercising that right, cyclists must give way to pedestrians and horseriders. The right can be taken away by a Traffic Regulation Order, under the procedures described in the previous chapter, or under local byelaws.

The right conferred on cyclists is not to place any additional obligation on the Highway Authority with regard to the maintenance of the bridleway. Authorities do not, therefore, have to do anything to make the surface of bridleways more comfortable or safer in order to facilitate their use by cyclists. However, a cyclist on a mountain bike should have no difficulty in taking advantage of section 30 to make use of the bridleway network.

### Dedication of new footpaths or bridleways

Under section 25 of the Highways Act 1980, either a County or District Council may enter into an agreement with any landowner for the dedication of a footpath or bridleway over his land. An agreement under this section shall be on such terms as to payment or otherwise as may be specified in the agreement and may, if agreed between the Council and the landowner, provide for the dedication of the footpath or bridleway subject to limitations or conditions affecting the public right of way over it.

Upon the dedication of a footpath or bridleway, under a Section 25

Agreement, the Highway Authority shall survey the way
what work is necessary to bring it into a fit condition for us
public. If the Highway Authority has itself entered in.
agreement with the landowner, then they meet the costs of carry..
out the necessary works. If a District Council has entered into the
dedication agreement with the landowner, then it is for them to meet
the costs of the necessary works to make the way fit for public use.
Once these works have been paid for, the responsibility for
maintaining the footpath or bridleway thereafter falls upon the
Highway Authority.

Compulsory powers to create footpaths and bridleways are
available under section 26 of the Highways Act 1980. A Highway
Authority or a District Council can make a Public Path Creation
Order creating a footpath or bridleway over land. If the order is
unopposed, it can be confirmed by the council which made it.
Otherwise it has to be submitted to the Secretary of State and he
will hold a local inquiry if there are objections before deciding
whether or not to confirm the order. Again, the council which made
the order will have to pay for making the path fit for public use and
thereafter the responsibility for maintenance will be that of the
Highway Authority. If a landowner can show that the value of his
land has been depreciated by the making of the order, or a person
has suffered damage by being disturbed in his enjoyment of land
through the making of the order, compensation will be payable
under section 28 of the Act by the Authority by whom the order was
made equal to the amount of the depreciation or damage.[1]

These powers to create new footpaths and bridleways by agreement
or by compulsory order are not widely used because there is already
an enormous network of rights of way in existence. However, the
Countryside Commission and local authorities do work together in

---

1  The section is probably intended to confine the right to compensation to owners or
occupiers of land over which a path is created. It does not expressly say this but subsection
(4) provides that compensation is only payable if the creation of the path would have been
actionable, were it not done under statutory powers. It is possible that an owner could
succeed in an action for nuisance to the enjoyment of his property if his neighbour allowed
a new public path to be created close to the boundary between the two properties. The
Judge touched on this compensation aspect, without deciding the point, in *Allen v. Bagshot
R.D.C.* 1970, discussed in the next chapter.

naking use of them in order to create the long distance routes envisaged under the National Parks and Access to the Countryside Act 1949, described earlier in this chapter.

## The Wildlife and Countryside Act 1981

*Definitive Maps of rights of way*
The law relating to Definitive Maps is now contained in sections 53 to 57 of the Wildlife and Countryside Act 1981.

The County Councils and the Metropolitan District Councils are the Surveying Authorities for the purposes of the Act and must prepare and keep a Definitive Map and Statement showing the rights of way in their area. This map and statement must be open to public inspection, free of charge, at all reasonable hours. In the case of the Counties, copies of the map and statement must also be available for public inspection in each District of the County. This is usually achieved by letting the District Councils have copies. So far as appears practicable, copies should also be made available in each parish, but the Surveying Authority can comply with that requirement by letting the Parish Council have a copy of so much of the map and statement as relate to the parish.

The only exception to this obligation to prepare maps and statements of rights of way covering the whole of England and Wales is in London. The Inner London Boroughs do not have to prepare these maps and statements unless they resolve to adopt sections 53 to 57 within their Boroughs.

A Definitive Map and Statement are 'conclusive evidence' as to the particulars contained therein. This means that where, for example, the map shows a footpath, it is not open to a farmer charged with obstructing it to argue that the path is not a right of way at all. In the same way as the map is conclusive evidence of the existence of a right of way shown on it, any particulars in the statement are conclusive evidence as to its position and width.

The fact that the map shows a footpath is not, however, conclusive evidence that there is not also a right for horseriders over it. If it shows a bridleway, this is not conclusive evidence that there is not also a right for vehicular traffic.

174

Notwithstanding the fact that a right of way is shown as being a byway open to all traffic, it is still open to the Highway Authority to make a Traffic Regulation Order restricting the use of the way to pedestrians and horseriders, if they so wish. Also, under section 54(7), Parliament has made it clear that the Highway Authority is not obliged to provide a metalled carriageway on a byway open to all traffic.

## Continuous review of Definitive Maps and Statements

Under the 1949 Act, there were to be five yearly regular reviews of the whole map. Few Surveying Authorities managed to achieve this. They are no longer required to review the whole map at one time.

Instead, they must, under section 53 of the 1981 Act, keep their Definitive Map and Statement under continuous review. They must, as soon as is reasonably practicable after the occurrence of certain specified events, make modifications to the map and statement by formal orders in consequence of the occurrence of those events. Some of the events would obviously require modification of the map whilst others, as will be seen, can give rise to a great deal of argument.

The events which require an order modifying a Definitive Map to be made are specified in subsection (3) of section 53. They fall into three classes:

(a) the coming into force of an event such as an order creating, stopping up or diverting a highway including orders under sections 26, 116, 118 and 119 of the Highways Act 1980, and agreements with landowners under section 25 of the 1980 Act;[1]

(b) the end of the twenty year period which gives rise to a presumed dedication of a right of way;[2]

(c) the discovery of evidence showing that a right of way not shown on the map should be shown, or that what is shown on the map as a right of way should not be shown, or that what

---

1 See next chapter as to stopping up and diversion orders.

2 See Chapter 2 as to the presumption of dedication after twenty years' use.

is shown on the map as a right of way should be shown as a right of way of another class.

A modification order in respect of class (a) above is made automatically. This is because the order or other event leading to the modification will itself have been subject to statutory procedures, such as objections and a local inquiry, before the order was made. Modification orders in respect of class (b) or (c) above will be subject to the procedural requirements set out in Schedule 15 to the 1981 Act and discussed below.

Many of the matters contained in (b) and (c) will be contentious. A landowner may not agree that a path across his land has been used as of right over the last twenty years and may argue that it has been used by his permission. An old map or inclosure award may come to light which appears to show that a path was a public right of way. Nevertheless, the interpretation of the award could be arguable or a landowner who is affected by the discovery of the way may argue that it was lawfully stopped up or diverted in times past and that the formal order has been lost. An old Stopping Up Order may come to light which shows that a particular right of way should not have been included in the map.

**Procedure for making modifications**

Orders making modifications are prepared by the Surveying Authority and must then be publicised by them to allow an opportunity for objections to be made to the proposed modification (other than the orders described above, which simply give effect in the Definitive Map to creation agreements and orders, and stopping up and diversion orders, which have already gone through other statutory processes). The procedure is set out in Schedule 15 of the 1981 Act. Under it, notice of the proposal has to be given in a local newspaper and has to be served on every owner and occupier of land affected. Notice also has to be displayed in a prominent position at the ends of so much of any way[1] as is affected by the order. A period of six weeks has to be allowed for objections.

---

1  As to the meaning of "ends of the way" see *Ramblers' Association v. Kent C.C.* 1990 discussed in the next chapter.

If no objections are made, the Surveying Authority may confirm the order as an unopposed order and alter the map and statement. If there are objections, the order has to be submitted to the Secretary of State for confirmation by him. He will usually arrange for a local inquiry to be held, but he need not do this so long as he affords to any person by whom a representation or objection has been made, an opportunity of being heard by a person appointed by him.

Decisions on opposed orders are normally made by the Inspector holding the inquiry unless the Secretary of State directs that the decision is to be made by him following the inquiry.

Once the final decision has been made, notice of confirmation of an unopposed order, or notice of confirmation, with or without modification, of an opposed order by the Secretary of State, has to be given in the local press, posted on the right of way concerned and served on every owner and occupier of the land affected. Any person aggrieved by an order making modifications to the Definitive Map and Statement can, within six weeks of the public notice having been given of confirmation of the order, apply to the High Court to quash the order if he can show that the order made is not within the powers of the Surveying Authority or the Secretary of State, or that his interests have been substantially prejudiced by a failure to comply with the procedural requirements in Schedule 15. Except on these grounds, the validity of an order cannot be questioned in any legal proceedings.

Where new documentary evidence has come to light and has prompted the Surveying Authority to make the Modification Order, any person objecting to a proposed order can require the Surveying Authority to inform him what documents were taken into account in preparing the order. If the documents are in the possession of the Authority, he must be allowed to inspect them and take copies. If they are not in the possession of the Authority, they must give him any information they have as to where the documents can be inspected.

## Applications to Surveying Authorities for Modification Orders

Modification Orders can only be proposed and advertised by

Surveying Authorities. It is not uncommon for bodies, such as the Ramblers' Association or the British Horse Society, to consider that a new right of way has come into existence through twenty years' uninterrupted use as of right. Equally, a landowner may think that a right of way has been wrongly included in the Definitive Map or that it has been given a higher classification than it should have been, such as a byway instead of a bridleway, and he may be able to uncover new evidence which appears to show that he is right. There is, therefore, a procedure laid down in Schedule 14 of the Act for applications to be made by any person to the Surveying Authority for them to make a Modification Order.

Such an application has to be accompanied by a map showing the way to which the application relates and by copies of any documentary evidence and statements of witnesses which the applicant considers will support his case. The applicant must notify every owner and occupier of land affected by the application.

The Surveying Authority is under a duty to investigate the matter stated in the application and to consult with every other local authority whose area includes the land to which the application relates. The Surveying Authority must then decide whether or not to make the Modification Order for which the application is being made. The investigations of the Surveying Authority may take some time, particularly, as is quite often the case, if the evidence as to use is hotly disputed. An applicant may argue that a path has been used as of right for the last twenty years, whilst the landowner may consider that it is only through his permission that the public have been allowed to walk or ride the route.

The Surveying Authority are, therefore, allowed twelve months in which to make up their minds on the application. If they have not reached a decision by the end of twelve months, the applicant can ask the Secretary of State for the Environment to fix a further period within which they must reach a decision.

If the Authority decide not to make an order, the applicant may, within one month of being told of that decision, appeal to the Secretary of State for the Environment. The Secretary of State shall then consider the material submitted with the application and has the power, if he thinks fit, to direct the Surveying Authority to make the Modification Order requested.

Once a Modification Order has been made in response to an application, the procedure described above under Schedule 15 applies. The Authority will have to give notice of the making of the order and there will be a local inquiry if there are objections, followed by a decision of the Inspector holding the inquiry as to whether or not the modification should be made to the map.

## The conclusiveness of Definitive Maps – discovery of new evidence

Subsections (2)(b) and (3)(c) of section 53 of the 1981 Act require the Surveying Authority to consider evidence which becomes available to them after they have prepared the Definitive Map and Statement and to make such Modification Orders as appear to them to be necessary in the light of the evidence. These provisions have received detailed consideration from the Courts and in particular in two cases heard in 1989.

In the first case,[1] Mr. Riley, who was a keen motor-cyclist, asked the Wiltshire County Council to reclassify two bridleways as byways open to all traffic. The County Council refused to make a Modification Order and the Secretary of State refused to direct them to make an order. Mr. Riley had made a previous attempt to have these ways classified as byways, but at the time of his earlier application he had not only to show that there was a vehicular right of way over the track but also that it was suitable for vehicular traffic and that the extinguishment of vehicular rights of way would cause undue hardship. He succeeded in the earlier application in showing that there had been a vehicular right of way but was unable to show that the right of way was suitable for vehicular traffic or that the extinguishment of vehicular rights of way would cause undue hardship.

When Mr. Riley made his second application for reclassification of the ways, the law had changed and it was only necessary for him to

---

1   *R. v. Secretary of State for the Environment ex p. Riley* 1989.

establish that a public right of way for vehicular traffic had been shown to exist.

The County Council and the Secretary of State rejected the second application for a Modification Order because they said that Mr. Riley had already proved on the first application that there was a vehicular right of way and:

> "The additional evidence supplied did not really add to the weight of the evidence previously considered when it was acknowledged that the route was that of a former Turnpike Road, which became a County responsibility until it was dismained in 1893."

However, the High Court held that, even though it had been accepted on the earlier application that a vehicular right of way did exist, Mr. Riley had produced further evidence in support of that contention. The Surveying Authority must consider that further evidence and, now that the other two grounds for classifying a way as a byway no longer had to be established, they must upgrade the two bridleways to byway status.

The second case in 1989[1] came before the Court of Appeal and related to two different applications, one in Buckinghamshire and the other in Leicestershire. In the Buckinghamshire case, the appellants asked for two bridleways to be deleted from the Definitive Map altogether and in the Leicestershire case the appellant claimed that a bridleway had been incorrectly recorded and should have been described as a footpath. Both cases raised the same issue – should the Surveying Authority consider evidence coming to light after the Definitive Map was prepared but relating to the status of the way before it had been prepared; or do the provisions of section 56 of the 1981 Act, that a Definitive Map is conclusive evidence as the particulars shown on the map, preclude the consideration of fresh evidence?

The three Appeal Court Judges decided unanimously that Parliament had intended that fresh evidence about the status of a way should be considered irrespective of the period to which it

---

1  *R. v. Secretary of State for the Environment ex p. Simms and Burrows* 1989.

related and whether it could have been discovered before the map was prepared. Otherwise, cases could arise, and indeed had arisen, where new evidence showed beyond doubt that a mistake had been made in the preparation of the Definitive Map but nevertheless the map could not be corrected.[1]

The Definitive Maps are, therefore, conclusive evidence of the rights of way shown on them but only in the absence of evidence to the contrary. However, changes to the maps can only be made by going through the Modification Order procedure, as explained above. This requires the giving of notice of any proposals to modify a map and the making available to objectors of any new documentary evidence which has come to light. If objectors feel that the evidence does not justify modifying the map, they can maintain their objections so that the Modification Order cannot be confirmed as an unconfirmed order and so that there will be a local inquiry in which the new evidence can be challenged or countered by other evidence produced by the objectors.

The conclusive nature of the maps therefore remains of enormous value in enforcing the legislation regarding the protection of rights of way. An occupier of land, faced with a prosecution for obstructing a path or erecting a deterrent notice next to it, cannot argue before the Magistrates that the path should not have been shown as a right of way on the Definitive Map. Until new evidence has been produced bearing on its status, and has been through the Modification Order procedure, the path will be treated as a right of way and the occupier will not have a defence to a prosecution for, say, obstructing it. Equally, a Highway Authority cannot shirk their duty of erecting a sign showing the path where it leaves a metalled road while it remains on the Definitive Map.

## Bulls

Under section 59 of the 1981 Act, the occupier of a field or inclosure crossed by a right of way must not permit a bull to be at large in the field or inclosure. The section does not apply to bulls under the age of ten months. There is also an exception for bulls which are not of

---

1 *Suffolk C.C. v. Mason* 1979.

a recognised dairy breed and which are at large in a field or inclosure in which cows or heifers are also at large. Failure by an occupier to comply with section 59 is an offence punishable on conviction by a fine of up to one thousand pounds.

## Rights of Way Act 1990

*Damage to the surface of paths, ploughing out of paths and interference by growing crops*
The enactment by Parliament of the Rights of Way Act in 1990 was a further step to protect the existence of rights of way on the ground. A new section, 131A, was included in the 1980 Highways Act to make it explicit that any person who disturbs the surface of a right of way so as to render it inconvenient for the exercise of the right of way is guilty of an offence.

Farmers sometimes plough out a right of way. The right of way could have been dedicated subject to a condition that the farmer reserved the right to plough it.[1] Limited dedications were discussed in Chapter 2, and section 3 of the Rights of Way Act 1990 specifically protects such a right. Most farmers will have difficulty in showing that a path dedicated long ago was subject to such a condition concerning ploughing. The Highway Acts have therefore authorised farmers to plough paths providing they reinstate them after the ploughing has been completed.

The arrangements for this were embodied in section 134 of the Highways Act 1980. A new section 134 has been substituted for the old one by the 1990 Act, making the requirement for restoring the path properly and within a specified time much more stringent. The farmer has fourteen days from the first disturbance for the purposes of sowing a particular crop in which to make good the surface of the path. If the path is again disturbed during the course of the year, the farmer has only twenty-four hours within which to reinstate the path. As well as restoring the path after ploughing, the farmer must also indicate the line of the path on the ground so that it is apparent to members of the public wishing to use it.

---

1  *Mercer v. Woodgate* 1869.

The Act also, for the first time, lays down minimum widths for rights of way and provides that reinstatement is to be carried out to those minimum widths. In a case when the width of the right of way can be proved, the minimum width is the proved width. In other cases the minimum widths are one metre for a footpath crossing a field and one-and-a-half metres for a footpath around the edge of a field. The minimum width for a bridleway which is not a field edge path is two metres, and the width for a byway is three metres.

The provisions for Highway Authorities to authorise other works, apart from ploughing, which will have the effect of disturbing a footpath or bridleway on agricultural land were also made stricter by the 1990 Act, by substituting a new section 135 in the 1980 Act. The occupier of agricultural land can apply to the Authority to make an order for a period of up to three months, authorising a temporary diversion of the path and such an order may include such conditions as the Authority think fit for the provision of facilities, either by the applicant, or by the Authority at the applicant's expense, for the convenient use of any such diversion, including signposts, notices, stiles, bridges and gates. The order may also include conditions for making good the surface of a path to its minimum width before the expiration of the three month period.

If the applicant fails to comply with a condition on an authorisation, such as the erection of signposts or making good the surface of a path within the authorisation period, he will be guilty of an offence and liable to a fine on conviction not exceeding one thousand pounds.

The Act has also inserted a new section 137A in the Highways Act 1980, providing that where a crop, other than grass, has been sown on agricultural land, the occupier of the land shall take such steps as may be necessary to ensure that the line on the ground of any path is indicated to not less than this minimum width, so as to be apparent to members of the public. The occupier must also take steps to prevent the crop from encroaching on the path so as to render it inconvenient for the exercise of a public right of way, for example by reducing the width of the path to less than its minimum width.

### Power to enter on land and carry out works

Prosecutions take time to complete and fines may not be a deterrent to certain owners and occupiers. The 1990 Act, therefore, contains valuable additional powers for Highway Authorities to enter on agricultural land where occupiers have failed to comply with the statutory arrangements described above. If a farmer has not reinstated a path after ploughing it, or after a temporary diversion, or if an insufficient width of path has been left between growing crops, the Highway Authority can enter upon the land and carry out such works as may be necessary to rectify the default. They can then recover the costs of carrying out the works from the occupier of the land who has failed to do them himself.

### Duty to protect rights of way

Reference was made in Chapter 5 to the express duty laid on the Highway Authority by section 130 of the Highways Act 1980 "to assert and protect the rights of the public to the use and enjoyment of any highway for which they are the Highway Authority". It is noticeable that in each of the sections which Parliament has either added or substituted in the 1980 Act by means of the 1990 Act, they have said that it is the duty of the Highway Authority to enforce the provisions of the new or substituted sections. This will be quite a heavy burden for Highway Authorities because, for instance, a large number of rights of way are ploughed out during each planting season and it will be the duty of the Authority to see that they are properly reinstated to the minimum width within the two week period. However, there is little doubt that the bodies concerned with the interests of walkers and riders will be watching Authorities closely to ensure that they do just that.

### Gates and stiles

A gate or stile is a physical obstruction of the highway. Section 137 of the 1980 Act, imposing a penalty for wilful obstruction, applies just as much to the erection of a gate or stile on a footpath or bridleway without lawful authority as it does to the obstruction of a metalled highway by way of a gate, or any other means. However, many footpaths and bridleways have been dedicated subject to the existence of gates and stiles across them and this is perfectly valid. Also, under section 147 of the 1980 Act, the owner, lessee or occupier of agricultural land may apply to a Highway Authority for permission for the erection of stiles, gates or other works for preventing the escape of animals. The Highway Authority may accept such an application and authorise the erection of stiles, gates or other works.

If the right of way was dedicated subject to the existence of gates and stiles, the duty to maintain the stiles and gates remains with the owner of the land and he must keep them in a safe condition and to the standard of repair required to prevent unreasonable interference with the rights of the persons using the footpath or bridleway. This is an exception to the normal rule that the Highway Authority is responsible for the maintenance of the highway and anything in it. A Highway Authority in giving consent under section 147 will also normally make it a condition of the consent that the occupier of the agricultural land shall be responsible for maintaining any gates and stiles to which they are consenting.

Unless a way has been dedicated subject to the existence of a gate, or the Highway Authority has authorised the erection of a gate, gates tied or looped together by string, even though the string can be untied by users of the way, constitute an obstruction for which the person responsible can be convicted under section 137 of the 1980 Act.[1]

## Maintenance of footpaths and bridleways

A Parish Council has power to undertake the maintenance of any footpath or bridleway within its parish without prejudice to the duty of the Highway Authority to maintain such a footpath or bridleway. The Authority may undertake to reimburse to the Parish Council the whole or any part of expenditure incurred by them in maintaining footpaths or bridleways. Owing to the size of the rural counties and the enormous network of rights of way, some Authorities have found that these provisions do enable them to maintain footpaths and bridleways more effectively by paying Parish Councils to deal

---

1 *Durham C.C. v. Scott* 1990.

with these responsibilities locally, rather than trying to cope with all the work from the centre.

The duty of the Highway Authority to maintain the highway applies just as much to rights of way as to roads. They will be liable in damages to persons sustaining accidents while using the rights of way if they have not fulfilled their duty. In the event of an accident, they may be able to rely on the special defence in section 58 of the 1980 Act (discussed in Chapter 4) that they had taken such care as in all the circumstances was reasonably required to secure that the right of way was not dangerous for users. These principles were illustrated in the case of *Whiting v. Hillingdon London Borough Council* (1970).

Mrs. Whiting was walking along a narrow, unsurfaced public footpath. She wanted to overtake some men walking ahead of her and they stood aside for her. As she was passing the last of the four men, she turned sideways to get past but, even so, owing to the limited space available, she trod in the grass and foliage along the edge of the footpath and her leg came into contact with the stump of a tree which had been felled. She sustained quite serious injuries.

The evidence of the Highway Authority's own officers was that the stump was dangerous and that, had they been aware of its existence, they would have taken action to have it removed. Much, therefore, turned on the Authority's system of inspection of rights of way and the frequency of inspections. Repairs had, in fact, been carried out to this particular footpath in February 1966, only two months before Mrs. Whiting's accident in April. The Authority was responsible for the maintenance of three hundred and eighty miles of footpaths in the Borough and inspected these only once a year. It was not necessary for the Judge to express a view on the adequacy or otherwise of an annual inspection because the accident had occurred only two months after the repairs had been carried out. He did say:

> "I accept the contentions put forward for the plaintiff that the local authority were not entitled to treat every footpath within their area alike as having equal importance, and I accept, on the evidence before me, that this was a footpath, whatever its origin, which had

become part of an area which had been developed both as to offices and residences and was a footpath that was well used. Therefore, it should be given greater priority than those other footpaths across open country, which have been referred to briefly in the evidence, not having the same degree of use as this one did."

The Judge was impressed by the Authority's officers and felt certain that, had the stump been on the path in February, they would have noticed it and done something about it. He, therefore, concluded that the tree had been felled by, or on behalf of, the adjoining occupier, and the stump left close to the path during the two month period between the repairs being carried out and the accident occurring. He did comment that, had the stump been there in the previous summer, when the path had its annual inspection, the Inspector might well not have noticed it because it would have been covered with foliage at that time of year. He went on:

"There was nothing that would draw his attention to it. I do not think that it was incumbent upon him to beat about in the foliage on the verges of a footpath proper, and at the root of that stump."

He concluded his Judgment by saying:

"The ultimate question, therefore, is:

Ought there to have been a further inspection after the repair in February 1966 before the date of this accident on April 21st of that year? In my judgment it would be asking too much of the local authority, having had a special inspection in December 1965 and not having notice of any danger of this sort being present at that time, because it was not there, and having carried out repair work in February 1966, having no notice of danger of this kind because it was not there at that time, to have had a further inspection between that date and the date of the accident. I find they were not in breach of their duty under the Highways Act 1959 and have established a defence under the Act of 1961: they took all reasonable care by all practicable steps to ensure that

187

this footpath was not dangerous for persons who might be likely and reasonably expected to use it."

Whilst, therefore, this case does not decide how often Highway Authorities should inspect footpaths, it does give some helpful pointers.

Chapter 12

# Diversion and Extinguishment
# of the Highway

At common law, there is a simple maxim – "once a highway, always a highway". Even though a highway has fallen into disuse its status as a highway remains.[1] Whilst a highway may be extinguished by natural causes, such as erosion by the sea or landslips,[2] even then a question of fact may arise as to whether the foundation of the highway has been completely obliterated.[3] The Courts will still ask whether the cost of repair is out of reasonable proportion to the value of the road before accepting that the highway has disappeared.

The only way of stopping up a highway in the Middle Ages was by a writ of Ad Quod Damnum. This was a writ addressed to the Sheriff of the County directing him to summon a jury to enquire whether a proposed extinguishment of a highway would be detrimental to the public. If the jury found in the negative, the Crown might grant a licence authorising the highway to be stopped up. In the absence of any general statutory procedure for extinguishing a highway, it was, until the middle of the nineteenth century, as difficult to remove highway rights as to obtain a divorce. Both required a private Act of Parliament.

Given the enormous difficulty of extinguishing highway rights, once established, it is not surprising that there have been so many cases down the centuries where landowners have disputed in the Courts that a right of way has been dedicated by them or their predecessors and some of these cases were discussed in Chapter 2.

---

1  *R. v. Platts* 1880; *St. Ives Corporation v. Wadsworth* 1908.

2  *R. v. Inhabitants of Paul* 1840.

3  *R. v. Inhabitants of Greenhow* 1876.

189

There are now statutory procedures for stopping up or diverting highways. Even today however, those seeking to stop up or divert a highway, including Government Departments, are likely to face an uphill task and cannot assume that, even with the aid of the statutory procedures, it will be easy to achieve the desired order. Not many years ago, the Government encountered resistance in an application to divert a footpath passing through the grounds of Chequers, the Prime Minister's country residence in Buckinghamshire.[1]

Although there are a number of statutory procedures, there are two points common to all of them:

1. Proposals to stop up or divert have to be made known to the public, usually by an advertisement in the press and by posting notices on the rights of way affected.

2. All Statutory Undertakers with apparatus in the highway affected by the proposals must be given notice of them. This is because under the street works procedures described in Chapter 9, they have rights to lay services under the highway and to maintain them. These rights may become much more difficult to exercise once the status of the land as a highway is extinguished. Nevertheless the undertakers retain the rights to maintain and repair their apparatus even though it is no longer under a highway. If reasonably requested to do so by the Highway Authority they must remove their apparatus to another position but this is to be at the expense of the Authority.

The procedures for stopping up or diverting bridleways and footpaths are slightly different from those for highways with vehicular rights and there is power for the Highway Authorities to confirm unopposed orders. However, orders affecting footpaths and bridleways are sometimes the most hotly contested by bodies such as the Ramblers' Association and the Open Spaces Society. The use by the public of rights of way in the countryside for recreational purposes continues to increase steadily and the network of footpaths and bridleways is an important national leisure resource. It is

---

1 *Roberton v. Secretary of State for the Environment* 1975.

noticeable that this network is not available in many other countries, including Scotland, where there has not been a strong tradition of the protection of highway rights through the legal system.

## Highways Act 1980 – section 116

Section 116 of the Highways Act 1980 provides:

> "... if it appears to a Magistrates' Court, after a view, if the Court thinks fit, by any two or more of the Justices composing the Court, that a highway ...
>
> (a) is unnecessary, or
>
> (b) can be diverted so as to make it nearer or more commodious to the public,
>
> the Court may by Order authorise it to be stopped up or, as the case may be, to be so diverted."

It will be noted that this procedure can only be initiated by the Highway Authority whereas in the alternative procedure available under the Town and Country Planning legislation, discussed later in this chapter, anyone can make an application for a stopping up or diversion order.

Under section 116, notice has to be given to the relevant District Council and Parish Council, and, if either of them withhold their consent, the Highway Authority cannot proceed with the application.

Notice has to be given to owners and occupiers of adjoining lands, to any Statutory Undertakers with apparatus under the highway, in the London Gazette, and in at least one local newspaper. Notice also has to be displayed "in a prominent position at the ends of the highway".

On the hearing of the application by the Magistrates, any person on whom notice is required to be served, and any person who uses the highway, and any other person who would be aggrieved by the making of the order, have a right to be heard. If the Magistrates make the order applied for, any of these persons, even if they did not appear before the Magistrates, have a right of appeal to the Crown Court against the making of the order.

If the application is for a diversion, an order cannot be made by the Magistrates before the written consent of every person having a legal interest in the land over which the highway is to be diverted is deposited with the Court. Also, in the case of a diversion, the order cannot authorise the stopping up of any part of the existing highway until the new part to be substituted has been completed to the satisfaction of two of the Magistrates.

In making an order for the stopping up or diversion of a highway under section 116, the Magistrates can reserve a right of way on foot or on horseback so that the highway rights are not completely lost.

The section authorises a diversion if the new route is more commodious to the public and this can mean drier, less liable to flooding, less likely to be periodically ploughed up or freer from gates or stiles, even though the alternative route may be longer.

Valuable guidance to Magistrates on how to apply the various provisions of section 116, set out above, was given by Lord Justice Woolf in the course of his judgment in the Divisional Court in the case of *The Ramblers' Association v. Kent County Council* (1990). In that case, questions arose both as to the validity of the notices given of the making of the application and as to the meaning of "unnecessary". So far as the notices were concerned, the Court held that the requirement to display notices "at the ends of the highway" meant at the ends of the length of highway to be diverted or stopped up. If a highway running between two other highways was to be stopped up or diverted for part of its length, it was not good enough to place notices of the application at the junction of the highway with the other two highways. The notices had to be placed at the points where the highway would be stopped up or diverted.

The notices were also wrong in referring to a diversion as well as to stopping up. The route to which part of the highway was to be diverted was only being made available by the landowner on a permissive basis and was not being dedicated as a highway in perpetuity. The Court held that the reference to a diversion was

misleading to such an extent as to deprive the Magistrates of the
right to make the orders sought.

On the question of the meaning of "unnecessary", Lord Justice
Woolf said:

> "I equally have little doubt that Magistrates, on the
> whole, are best left to determine what is unnecessary
> themselves. The very fact that there is an express
> reference, which is unusual, to the Magistrates making
> a view, indicates how much this is a question of fact and
> one in relation to which one would expect the
> Magistrates to use their local knowledge and common
> sense in coming to a decision.

> "However, it may provide some assistance to
> Magistrates in the difficult adjudicating task they have
> to perform under section 116(1) if I give the following
> guidance. First of all I consider that Magistrates, in
> deciding whether or not a highway is unnecessary,
> should bear in mind the question for whom the highway
> is unnecessary. It is to be unnecessary for the public. It
> is the public who have the right to travel up and down
> the way in question, and it is the public with whom the
> Justices should be concerned because the right is vested
> in them. It is for this reason that I drew attention to the
> somewhat different language in section 118.

> "Then the Justices might ask themselves, in considering
> an application under section 116, the question for what
> purpose should the way be unnecessary before they
> exercise their jurisdiction. So far as that is concerned, it
> should be unnecessary for the sort of purposes which the
> Justices would reasonably expect the public to use that
> particular way. Sometimes they will be using it to get
> primarily to a specific destination – possibly here the
> shore. Another reason for using a way of this sort can be
> recreational purposes.

> "In my view, where there is evidence of use of a way,
> *prima facie*, at any rate, it will be difficult for Justices

193

H

> properly to come to the conclusion that a way is unnecessary unless the public are or are going to be provided with a reasonably suitable alternative way."

On the question of whether the public would be given a reasonably suitable alternative way, the Court said that not only must the alternative route be dedicated as a highway but it must also be suitable for the purpose for which the public were using the existing way. The Magistrates must also consider whether the result of the loss of the way could make the other ways which were available more crowded.

If a way was being much used for recreational purposes, that would be a consideration which the Justices should take into account. They should not jump to the conclusion that the way was unnecessary because it was not necessary for people to use to get to and from work. The existence or otherwise of a suitable alternative route was of "critical importance" for the question of whether the existing route was unnecessary or not.

## Sections 118 and 119 of the Highways Act 1980

*Stopping up and diversion of footpaths and bridleways*
Sections 118 and 119 of the 1980 Act deal with the stopping up and diversion of footpaths and bridleways. There are three important differences between these two sections and section 116. First, on the question of stopping up, under section 118, the issue is not whether the path is unnecessary but whether it is expedient that the path should be stopped up on the ground that it is not needed for public use. Secondly, on a diversion, under section 119, the question is whether it is expedient that the line of the path should be diverted in the interests of the owner or occupier of the land rather than whether the new line is nearer or more commodious. Thirdly the order is made by the County or District Council not the Magistrates.

The use of the word 'expedient' might lead landowners, and any other persons supporting the stopping up or diversion, to think that they have a better chance of success than under the section 116 procedure before the Magistrates, described above. However, both sections are hedged about with strict controls. Stopping up under section 118 is not to be effected unless the Authority or the Secretary of State, as the case may be, are satisfied that it is expedient to stop up, having regard to the extent that the path would, apart from the order, be likely to be used by the public. Whilst, therefore, it might be expedient in the interests of the landowner to stop up the path and whilst the path may not be *needed* by the public, yet if it is much used, say for recreational purposes, then it should not be stopped up. On a Diversion Order, made under section 119, the Secretary of State or the council, as the case may be, are not to confirm a Diversion Order unless they are satisfied that the new path will be "substantially as convenient to the public" in respect of its starting and finishing points and that it is expedient to confirm the order having regard to the effect which the diversion would have "on public enjoyment of the path or way as a whole".

As regards diversions under section 119, a Diversion Order must not alter a point of termination of the path to a point not on a highway or, where it is on a highway, otherwise than to another point which is on the same highway, or a highway connected with it. Also, when works are needed to make the new line of the path fit for public use, the date in the order for extinguishing the existing line of the path must be later than the date when the new line is created in order to allow sufficient time for the works to be carried out.

Orders under section 118 and 119 have to be advertised in the local press and notified to every owner and occupier of the land crossed by the path. Copies of the notice also have to be displayed at the ends of the path to be stopped up or diverted.

It is worth noting also that the consent of a landowner over whose land a path is to be diverted is not required under section 119. This is in marked contrast to a section 116 application dealing with roads where the Justices cannot proceed to hear an application without the written consent of the landowners over whose lands the road is to be diverted.

Written notice of the order proposed does have to be served on all owners and occupiers of the land over which the path is to be diverted. They may well object to the new route over their land but the Secretary of State can confirm the order if he thinks fit after holding an inquiry. If the order is confirmed, the owners and occupiers of any land over which the path has been diverted will be entitled to compensation for the depreciation in the value of their land or for damage suffered by being disturbed in the enjoyment of their property. No doubt, for this reason, section 119(5) enables the local authority to make an agreement with the landowner who is seeking the diversion of the path away from his land, that he should defray or contribute towards the compensation which will have to be paid to the owner onto whose land the path is diverted.

If there are no objections, the council can confirm the order subject to the considerations set out above. If there are objections, the order must be referred to the Secretary of State for confirmation. He will usually appoint an Inspector to hold a local inquiry and to decide whether or not to confirm the order, but he can simply give to any person by whom an objection has been made the opportunity of being heard by a person appointed by him. The Secretary of State, or the Inspector if he has delegated powers, can confirm the order with or without modifications or refuse to confirm the order.

The Chequers case mentioned at the beginning of this chapter arose on a Diversion Order made under section 111 of the Highways Act 1959, in respect of a countryside footpath crossing fields. The order was made for the purpose of taking the path further from the house because the police were anxious about the ease with which a terrorist standing on the path could shoot at the Prime Minister and his guests sitting on the terrace of the house. The Secretary of State decided to confirm the order in spite of local objections and defended the decision before the High Court on the basis that the order was necessary to "secure the efficient use of land", this being the phrase used in the Highways Act 1959.

The Judge held that the expression "land" did not only mean the surface of the path, but the estate of which it was a part. He also found that "efficient use of land" was not confined to agricultural considerations. The diversion was necessary to enable the house to be used to its best advantage for a country residence for Prime Ministers, for which purpose it had been given to the nation in 1917. He agreed that on these particular facts it could be argued that the diversion was made to secure the efficient use of the land. It is likely that a Court would come to the same decision today, bearing in mind the wording in the 1980 Act "in the interests of the owner", so long as the diversion did not have a serious detrimental effect on the enjoyment by the public of the path as a whole.

The case of *Allen v. Bagshot Rural District Council* (1970) decided two points of interest on Section 119 Diversion Orders. In that case, a proposed Diversion Order re-routed the line of the path to a route alongside, but just outside, the boundary of Mr. Allen's house from its existing position some distance away and people walking on the new line of the path would overlook his kitchen and bathroom windows.

Although the notices of the proposed order were correctly erected on the path, and one was just outside the hedge along the boundary of Mr. Allen's property, neither he nor his wife did anything about it during the twenty-eight days for which the notice had to be displayed. Mr. Allen did not look at the notices carefully until three or four weeks after the expiry of the twenty-eight day period and did not object to the council until six weeks after the end of the objection period. By that time the order had been confirmed by the council as an unopposed order.

The Judge held that any representation or objection made outside the twenty-eight day period is not duly made and the council could confirm an order as an unopposed order, even if objections are received after the twenty-eight day period and before confirmation. Equally, the Judge said that the Secretary of State could have confirmed the order without holding a local inquiry if the objection was not made in due time.

The second point which the Judge decided was that Mr. and Mrs. Allen were not entitled to be individually served with notice of the proposed order because the line of the diversion would not run over their land, nor were they likely to be entitled to compensation. The Judge said:

> "Hard though it may seem for a person in the applicant's position to be told that he is not a person whose interests have to be considered when it is proposed to run a public

.ootpath right alongside his boundary, this is in accordance with the general principles of planning and compensation law. If it is a hardship to the applicant, it is a hardship which is shared with many other citizens."

## Acquisition of Land Act 1981 – section 32

An authority with compulsory purchase powers, for example a Government Department or a local authority, can make an Extinguishment Order in respect of any public footpath or bridleway over land which they have acquired if they are satisfied that a suitable alternative right of way has been or will be provided, or that the provision of an alternative way is not required. They must follow the same procedure for advertisement as under section 118 of the Highways Act 1980. If there are no objections, a local authority can confirm an unopposed order, but otherwise the confirming authority is the Secretary of State.

## Town and County Planning Act 1990 – section 247

*Stopping up by Secretary of State to enable development to be carried out*
Under section 247 of the Town and Country Planning Act 1990:

> "The Secretary of State may by Order authorise the stopping up or diversion of any highway *if he is satisfied that it is necessary to do so in order to enable development to be carried out*:
>
> (a) in accordance with planning permission granted under Part III, or
>
> (b) by a Government Department."

Part III deals with the grant of planning permission for development. If planning permission has been granted for new development, such as the construction of houses or factories, and the application site is crossed by a highway, whether a road or footpath, then an application can be made under section 247 to stop up or divert the road or footpath.

On the face of it, this appears to be an easier procedure than that under section 116 of the 1980 Act. An application to the Secretary of State can be made by anyone and there is no need to show that the highway is "*unnecessary*" or that an alternative route for a diversion is "*nearer or more commodious*". All that has to be demonstrated is that the development for which planning permission has been granted cannot be carried out unless the highway in question is stopped up or diverted.

Once again, however, there are strict requirements for publicising a proposal to make an order in a local newspaper and in the London Gazette. The Secretary of State has to serve a copy of the draft order on every local authority in whose area the highway is and on all the Statutory Undertakers who have apparatus under the highway. Copies of the notices have to be displayed at the ends of the highway to be stopped up or diverted. If objections are received, the Secretary of State will usually cause a local inquiry to be held before making the order.

There are two important points to note about this procedure under the Town and Country Planning legislation. First the Secretary of State has no power to make an order once the development has been completed. The Courts have stretched a point and ruled that there is power to make an order while the development is still proceeding if this is necessary to enable the remainder of the development to be carried out.[1] If, however, it emerges that, after the development is completed, there is a little used public footpath crossing the development site, it will be too late to use this procedure to stop up the right of way.

Secondly, in the recent case of *Vasiliou v. The Secretary of State for Transport* (1990), the Court of Appeal pointed out that no compensation was payable under section 247 to anyone who had suffered loss through the stopping up of the highway[2] and that the Secretary of State had a discretion as to whether or not to make an order even if it had been shown that the stopping up was necessary to enable the development to be carried out.

---

1  *Ashby v. Secretary of State for the Environment* 1979.

2  *Jolliffe v. Exeter Corporation* 1967.

Mr. Vasiliou owned a taverna in a street in Blackpool which was the subject of a stopping up application to the Secretary of State. Between 60% and 70% of his business was passing trade and if the road was stopped up the business would be likely to fail. The Inspector holding the local inquiry was impressed by the hardship which would be caused and recommended that a Closure Order should not be made. The Secretary of State rejected his Inspector's recommendation and said that the section was solely concerned with highway matters. He argued that Mr. Vasiliou should have objected to the planning application for the development which would necessitate the closure of the street. The Court said that financial loss to Mr. Vasiliou would not be a material consideration in deciding whether to grant planning permission or not. Even if it had been, Mr. Vasiliou was not restricted to objecting at that stage and the Inspector had been quite right to take account of his objections in considering the section 247 stopping up order.

The Court held that the Secretary of State ought to take into account the adverse affect his order would have on those entitled to any rights which would be extinguished by the order, especially since the statute made no provision for compensation for them. The Secretary of State had misdirected himself and the Court quashed the stopping up order.

As with the Highways Act powers, there is power to divert a way as well as to stop it up. Section 247 additionally provides that the Secretary of State can include in his order directions for "the provision or improvement of any other highway". The developer seeking the order is likely to be required to pay the costs of such work under subsection (4)(a)(i).

There appears to be no provision in section 247, or section 252 dealing with the procedure, for any person onto whose land the way is to be diverted to be notified of the proposed order, nor does there appear to be any provision for him to claim compensation. If he sees advertisements of the proposed order, or notices on the way to be diverted, he can object but the Secretary of State may decide, after an inquiry, to disregard his objection and make the order.

## Town and Country Planning Act 1990 – sections 257 and 258

*Stopping up by local planning authority to enable development to be carried out*

Section 257 of the Town and Country Planning Act 1990 deals with the stopping up or diversion of footpaths or bridleways. It enables the council, which granted a planning permission, to make a stopping up or diversion order in respect of any footpath or bridleway if it is satisfied that it is necessary to do so in order to enable the development to be carried out in accordance with the planning permission granted by it.

Schedule 14 of the 1990 Act applies to orders made under section 257 and requires the authority making the order to notify owners and occupiers of any land over which it is proposed to divert a path. However, there appears to be no provision for the payment of compensation to a person over whose property the path is to be diverted.

Section 258 enables a local authority which has acquired or appropriated land for planning purposes to make an order extinguishing any footpath or bridleway if they are satisfied that an alternative right of way will be provided or is not required.

The local authority must give notice of its intention to make an order under either of these two sections in the press and must display the notice in a prominent position at the ends of the footpath or bridleway to be stopped up or diverted. Notice has to be given to owners and occupiers of land crossed by the path, to every Parish Council and to all Statutory Undertakers who have apparatus under the path. If no objections are received to the order, the authority who made it can confirm it as an unopposed order. If there are objections, they must be referred to the Secretary of State who will either cause a local inquiry to be held or give any person who has objected an opportunity of being heard by a person appointed by the Secretary of State. The Secretary of State may then confirm the order with or without modifications or may refuse to confirm the order.

## Side Road Orders

Where major new roads are proposed to be constructed, such as a motorway or a dual carriageway, it is often necessary to stop up highways which would otherwise cross the new road. Alternatively, engineering works can be carried out to alter the level of roads or paths in order to take them over a major road by a bridge or through an underpass. Consequently, there is power in section 14 of the 1980 Act to authorise the local Highway Authority for a classified road to stop up, raise, lower or otherwise alter a highway that crosses or enters the route of the road. The authorisation is achieved by way of an order made by the the Secretary of State for Transport or made by the local Highway Authority and confirmed by the Secretary of State. An order can provide for the preservation of the rights of statutory undertakers in respect of any apparatus of theirs which is under the highway to be stopped up or diverted.

Motorways, as was explained in Chapter 1, are special roads and orders can also be made by the Secretary of State under section 18 of the 1980 Act to stop up, divert, raise, lower or otherwise alter a highway which crosses or enters the route of the special road. Section 248 of the Town and Country Planning Act 1990 contains similar powers for the Secretary of State to stop up or divert highways crossing new roads if it appears to him to be expedient to do so in the interests of the safety of users of the main highway or to facilitate the movement of traffic on the main highway.

These orders are known as Side Road Orders. Public notice must be given of the proposed orders in the press and by placing notices at the ends of the highway to be stopped up or diverted. Any Undertakers, or sewerage authorities with apparatus under a highway affected, have to be notified individually.

### Extinguishment of vehicular rights – pedestrian precincts

A procedure is available in section 249 of the Town and Country Planning Act 1990 to extinguish the right to use vehicles on a highway which would otherwise be a highway for all purposes. The local Planning Authority can resolve to improve the amenity of part of their area and can make an application to the Secretary of State for him to provide by order for the extinguishment of any right which persons may have to use vehicles on a highway in that part. Much the same result can be achieved by a Traffic Regulation Order, discussed in Chapter 10. An order under section 249 extinguishes the vehicular rights for all time, though the local Planning Authority can apply for the Secretary of State to revoke it. If a person does have access to a highway over which the vehicular rights are taken away under this power, he can obtain compensation from the local Planning Authority for any depreciation in the value of his property attributable to the making of the order.

## Ministry of Defence land

There are special powers in a number of Defence Acts for the Government to stop up or divert highways crossing land vested in the Secretary of State for Defence or crossing airfields used in connection with the manufacture of aircraft for defence purposes. Powers are contained in the Defence Act 1842, section 17, the Defence Act 1860, section 40, the Military Lands Act 1892, section 13, and most recently the Land Powers (Defence) Act 1958. Under section 8 of the 1958 Act, the Secretary of State for Transport can stop up or divert a highway either permanently or temporarily where land is used by the Government for the purposes of an installation provided for defence purposes or is used by a manufacturer of aircraft as an airfield in connection with defence purposes. The procedures follow those under section 247 of the Town and Country Planning Act 1990, discussed above, and advertisement of the proposals is required in the usual way.

Section 3 of the Manoeuvres Act 1958 enables local Justices to suspend rights of way during military exercises for periods of up to 48 hours. Seven days' notice of any application to the Justices for such an order has to be given in the local press.

## Exchange of land – straightening the boundaries of the highway

### Agreements with neighbouring landowners

Under section 256 of the Highways Act 1980, the Highway Authority for any highway may, for the purpose of straightening or otherwise adjusting the boundaries of the highway, enter into an agreement with the owner of any land which adjoins or lies near to the highway providing for the exchange of any such land for land on which the highway is situate, with or without the payment by either party of money for equality of exchange.

A Highway Authority proposing to enter into an agreement under the section has to publish notice in at least two successive weeks in the local press of the proposed agreement and serve a copy of the notice on any Statutory Undertakers appearing to be affected by the proposed agreement. A copy of the notice must also be displayed in a prominent position on the part of the highway to which the proposal relates.

If no objections are made within two months of the giving of the public notice, the Authority can enter into the proposed agreement. A person who wishes to object can appeal to a Magistrates' Court against the proposed agreement. The Magistrates, after considering representations made by any party to the appeal and the desirability in the public interest of the proposed agreement, can either dismiss the appeal or order the Authority not to enter into the agreement.

Equivalent areas of land do not have to be exchanged as there is provision for the payment of money by way of equality of exchange. The Authority can transfer to an adjoining land owner a greater or lesser area of land than it receives back.

If the procedure is satisfactorily completed, and objections are either not made or are dismissed, then once the agreement has been entered into with the adjoining owner, the land which formed part of the highway is freed from the public right of way over the land. However, the Statutory Undertakers have the same rights in respect of apparatus in the land as if the agreement had not been entered into. Under Schedule 12 of the 1980 Act the parties can come to an arrangement to relocate the apparatus in the new length of highway, or to provide other apparatus in substitution for the existing apparatus, at the expense of the Highway Authority.

The power to exchange highway land in section 256 is useful because it is available in cases not covered by section 116 of the 1980 Act or section 247 of the Town and Country Planning Act 1990. It is not necessary to show that the highway is unnecessary within the meaning of section 116 of the 1980 Act or that it needs to be stopped up in order to enable some new development to be carried out within the terms of section 247 of the 1990 Act.

# Index

J